Shelford Bidwell

Curiosities of Light and Sight

Shelford Bidwell

Curiosities of Light and Sight

ISBN/EAN: 9783337250638

Printed in Europe, USA, Canada, Australia, Japan

Cover: Foto ©berggeist007 / pixelio.de

More available books at **www.hansebooks.com**

CURIOSITIES OF
LIGHT AND SIGHT

BY

SHELFORD BIDWELL, M.A., LL.B., F.R.S.

WITH FIFTY ILLUSTRATIONS

LONDON:
SWAN SONNENSCHEIN & CO., LIMITED
PATERNOSTER SQUARE
1899

PREFACE.

The following chapters are based upon notes of several unconnected lectures addressed to audiences of very different classes in the theatres of the Royal Institution, the London Institution, the Leeds Philosophical and Literary Society, and Caius House, Battersea.

In preparing the notes for publication the matter has been re-arranged with the object of presenting it, as far as might be, in methodical order; additions and omissions have been freely made, and numerous diagrams, illustrative of the apparatus and experiments described, have been provided.

I do not know that any apology is needed for offering the collection as thus re-modelled to a larger public. Though the essays are, for the most part, of a popular and informal character, they touch upon a number of curious matters of which no readily accessible account has yet appeared, while, even in the most elementary parts, an attempt has been made to handle the subject with some degree of freshness.

The interesting subjective phenomena which are associated with the sense of vision do not appear to have received in this country the attention they deserve. This little book may perhaps be of some slight service in suggesting to experimentalists, both professional and amateur, an attractive field of research which has hitherto been only partially explored.

CONTENTS.

Chapter I.
Light and the Eye 1

Chapter II.
Colour and its Perception ... 39

Chapter III
Some Optical Defects of the Eye ... 84

Chapter IV.
Some Optical Illusions ... 130

Chapter V.
Curiosities of Vision 165

LIST OF DIAGRAMS.

FIG.		PAGE.
1.	Image of Slit and Spectrum	12
2.	Diagram of the Eye	24
3.	Abney's Colour-patch Apparatus	45
4.	Partially Intercepted Spectrum	49
5.	Stencil Cards	52
6.	Helmholtz's Curves of Colour Sensations	72
7.	König's Curves	73
8.	Stencil Card for Complementary Colours	77
9.	Another form	79
10.	Slide for Mixing any two Spectral Colours ...	80
11.	Refraction of Monochromatic Light by Lens	87
12.	Refraction of Dichromatic Light ...	89
13.	Narrow Spectrum as seen from a Distance ...	97
14.	Spectrum formed with V-shaped Slit... ...	103
15.	Bezold's Device for Demonstrating Non-achromatism of the Eye...	108
16.	Crossed Lines showing the Effect of Astigmatism	113
17.	Another Design showing the same ...	114
18.	Star-like Images of Luminous Points...	116
19.	Sutures of the Crystalline Lens	117
20.	Multiple Images of a Luminous Point ...	120
21.	The same, showing an increased number of Images	122
22.	The same when a Slit is held before the Eye ..	123
23.	Multiple Images of an Electric Lamp Filament	125
24.	The same seen through a Slit ... ... 126—128

LIST OF DIAGRAMS.

FIG.		PAGE.
25.	Illusion of Length	132
26.	Another form	135
27.	Another form	136
28.	Another form	137
29.	Another form	138
30.	Illusion of Inclination	143
31.	Zöllner's Lines	144
32.	Slide for showing Illusions of Motions	147
33.	Illusion of Motion	149
34.	Illusion of Luminosity	152
35.	Illusion of Colour	155
36.	Recurrent Vision demonstrated with a Vacuum Tube	176
37.	The same with a Rotating Disk	178
38.	Apparatus for showing Recurrent Vision with Spectral Colours	181
39.	Charpentier's "Dark Band"	187
40.	Charpentier's Effect shown with the Hand ...	189
41.	Multiple Dark Bands	192
42.	Temporary Insensitiveness of the Eye after Illumination	194
43.	Visual Sensations attending a Period of Illumination	199
44.	Benham's Artificial Spectrum Top ...	200
45.	Demonstration of Red Colour-borders ...	205
46.	Black and White Screens for the same ...	209
47.	Rotating Disk for the same	210
48.	Demonstration of Blue Colour-borders ...	215
49.	Disk for Experiments on the Origin of the Colour-borders	217
50.	Disk for the Subjective Transformation of Colours	224

CHAPTER I.

LIGHT AND THE EYE.

In the present scientific age every one knows that light is transmitted across space through the medium of the luminiferous ether. This ether fills the whole of the known universe, as far at least as the remotest star visible in the most powerful telescopes, and is often said to be possessed of properties of so paradoxical a character that their unreserved acceptance has always been a matter of considerable difficulty.

The ether is a thing of immeasurable tenuity, being many millions of times rarer than the most perfect vacuum of which we have any experience: it offers no sensible obstruction to the movements of the celestial bodies, and even the flimsiest of material substances can pass through it as if it were nothing. Yet we have been taught that this same ether is an elastic solid with a great degree of rigidity, its resistance to distortion being, in comparison with the density, nearly ten thousand million times greater than that of steel: thus was explained the prodigious speed with which it propagates transverse vibrations.

A few years ago, a distinguished leader in science endeavoured in the course of a lecture to illustrate these apparently incompatible properties with the aid of a large slab of Burgundy pitch. He showed that the pitch was hard and brittle, yet, as he said, a bullet laid upon the slab would, in the course of a few months, sink into and penetrate through it, the hard brittle mass being really a very viscous fluid. The ether, it was suggested, resembled the pitch in having the rigidity of a solid and yet gradually yielding; it was, in fact, a rigid solid for luminiferous vibrations executed in about a hundred-billionth part of a second, and at the same time highly mobile to bodies like the earth going through it at the rate of twenty miles in a second.

This illustration, felicitous as it is, would, however, scarcely avail to force conviction upon an unwilling mind, even if it were admitted that the period of an ether wave is necessarily no more than a hundred-billionth of a second or thereabouts, which is probably very far from the truth.

But, indeed, the elastic solid theory of the ether has failed to give a consistent explanation of some of the most important points in observational optics; and, in spite of the exalted position which it has held, it can now hardly be regarded as representing a physical reality. The famous researches of Hertz have established upon a secure experimental basis the hypothesis of Maxwell that light is an electro-magnetic phenomenon. Such electrical radiations as can be produced by suitable instruments are found to behave in exactly the same manner as those to which light is due. They travel through space with the same speed; they can be reflected, refracted, polarised, and made to exhibit interference effects. No fact in physics can be much more firmly established than that of the essential identity of light and electricity. It follows then that the displacements of the ether which constitute light-waves are not necessarily of the same gross mechanical nature as those which we see on the surface of water, or which occur in the air when sound is transmitted through it. The displacements which the ether undergoes are not mechanical—primarily at all events—but electrical. Every one knows what a simple mechanical displacement is. If we push aside the bob of a suspended pendulum, that is a mechanical displacement. But if we electrify a stick of sealing wax by rubbing it with flannel, the surrounding ether undergoes electric displacement, and no one understands what electric displacement really is. Ultimately, no doubt, it will turn out to be of a mechanical nature, but it is almost certainly not a simple bodily distortion such as is caused, for example, when one presses a jelly with the finger.

Since, then, it is no longer necessary to assume that the exceedingly rare and subtile ether is a jelly-like solid in order to account for the manner in which it transmits light, one of the most serious difficulties in the way of its acceptance is removed. It is true that nothing is definitely known concerning the mechanism which takes the place of the simple transverse vibrations formerly postulated, but every one will admit that it is far easier to believe in what we know nothing about than in what we know to be impossible.

All scientific men are in fact agreed in recognising the real and genuine existence throughout space of an ether capable, among other things, of transmitting at the speed of 186,000 miles per second disturbances which, whatever their precise nature, are of the kind which mathematicians are accustomed to call waves. How an ether wave is constituted will probably be known when we have found out exactly what electricity is: and that may be never.

The sensation of light results from the action of ether waves upon the organism of the eye, but the old belief that the sensation was primarily due to a series of mere mechanical impulses or beats, just as that of sound results from the mechanical impact of air-waves upon the drum of the ear, cannot any longer be upheld. The essential nature of the action exerted by ether waves is still undetermined, though many guesses at the truth have been hazarded. It may be electrical or it may be chemical; possibly it is both. Ether-waves, we know, are competent to bring about chemical changes, as in the familiar instance of the photographic processes; they can also produce electric phenomena, as, for example, when they fall upon a suitably prepared piece of selenium; but there is no evidence that they can exert any direct mechanical action of a vibratory character, and indeed it is barely conceivable that any portion of our organism should be adapted to take up vibrations of such enormous rapidity as those which characterise light-waves.

Of the multitude of ether-waves which traverse space it is only comparatively few that have the power of exciting the sensation of light. As regards limited range of sensibility there is a very close analogy between hearing and seeing. No sensation of sound (at least of continuous sound) is produced when air-waves beat upon our ears unless the rate of the successive impulses lies within certain definite limits. It is just so with vision. If ether-waves fall upon our eyes at a less rate than about 400 billions per second, or at a greater rate than 750 billions per second, no sensation of light is perceived. There is another and more generally convenient way of stating this fact. Since all waves found in the ether travel through space at exactly the same speed—186,000 miles a second—it follows that the length* of each of a series of homogeneous waves must be inversely proportional to their frequency, that is, to the rate at which they strike a fixed object, such as the eye. Instead, therefore, of specifying waves by their frequency we may equally well specify them by their length. Waves whose frequency is 400 billions per second have a length of about $\frac{1}{34000}$ inch, this being the one four hundred billionth part of 186,000 miles; and those whose frequency is 750 billions have a wave-length of $\frac{1}{64000}$ inch.

---

* It should be clearly understood that the length of each wave of a series is measured by the distance between the crests of two successive waves. The length of water-waves which break upon a sea shore is not the length along the crest of a single wave measured in a direction parallel to the shore, as the uninitiated are apt to suppose. The true wave-length, or distance from crest to crest of successive waves, can be well observed from the top of a cliff.

Waves, then, of a length greater than $\frac{1}{34000}$ inch or less than $\frac{1}{64000}$ inch have no effect upon our organs of vision.*

In relation to this important fact it will be convenient to refer to a familiar but very beautiful experiment — the formation of a spectrum. An electric lamp is enclosed in an iron lantern, having in its front an upright slit; from this slit there issues a narrow beam of white light, which is made up of rays of many different wave-lengths, all mixed up together. By causing the light to pass through a prism the mixed rays are sorted out side by side according to their several wave-lengths, forming a

---

* In practice, wave-lengths are expressed in ten-millionths of a millimetre. The wave-lengths of the lines A and H of the solar spectrum, which approximately coincide with the limits of visibility, are 7594 and 3968 ten-millionths of a millimetre.

broad, many-hued band or "spectrum" upon a white screen placed to receive it. (See Fig. 1.) To the visible rays of the longest wave-length is due the red colour on the extreme left. Waves of some-

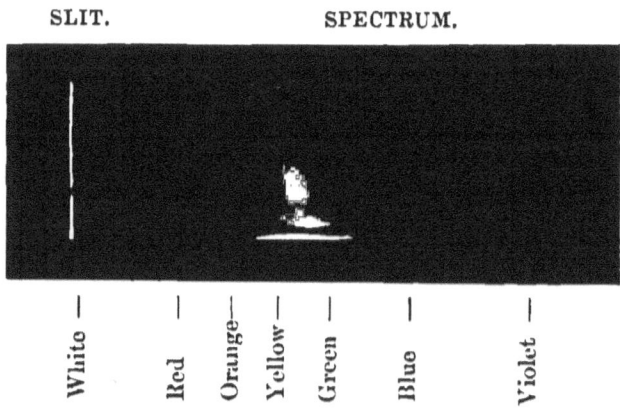

Fig. 1.—*Image of Slit and of Spectrum.*

what shorter length produce the adjoining stripe of orange, and the succeeding colours—yellow, green, and blue—correspond respectively to waves of shorter and shorter lengths. Lastly there comes a patch of violet due to those of the visible rays whose wave-length is the shortest of all. The wave-length of the light at the extreme edge of the red is about $\frac{1}{31600}$ inch, and as we pass along the spectrum the wave-length gradually diminishes, until at the extreme outer edge of the violet it is about $\frac{1}{61600}$ inch, or not much more than half that at the other end.

The two ends of the spectrum gradually fade away into darkness, and the point that I wish to insist upon and make perfectly clear is this:—The position of the boundaries terminating the visible spectrum does not depend upon anything whatever in the nature of light regarded as a physical phenomenon. Ether waves which are much longer and much shorter than those which illuminate the spectrum certainly exist, and evidence of their existence is easily obtainable. But we cannot see them; they fall upon our eyes without exciting the faintest sensation of light. The visible spectrum is limited solely by the physiological constitution of our organs of vision, and the fact that it begins and ends where it does is, from a physical point of view, a mere accident. The spectrum actually projected upon the screen is in truth much longer than that portion of it which any one can see: it extends for a considerable distance beyond the violet at the one end and beyond the red at the other, these invisible portions being known as the ultra-violet and infra-red regions. People's eyes differ in regard to range of sensibility just as their ears do. I believe the sensibility of my own eyes to be normal, but if I were to indicate the two points where the spectrum appears to me to begin and to end, a great many persons would certainly be inclined to disagree with me and place the boundaries somewhere else. Some, indeed, could see nothing whatever in what appears to most of us to be a brilliant portion of the red.

Again, it is by no means probable that in all animals and insects the limits of vision are the same as they are in man. We might naturally expect that larger and perhaps more coarsely constructed eyes than our own would respond to waves of greater average length, while the visual organs of small insects might on the other hand be more sensitive to shorter waves. The point is not one that can be easily settled, because we are unable to cross-examine an animal as to what it sees under different conditions. But Sir John Lubbock, taking advantage of the dislike which ants when in their nests have for light, has proved by a series of very exhaustive and conclusive experiments that these insects are most sensitive to rays which our own eyes cannot perceive at all. That region of the spectrum which appears brightest to the eye of an ant is what we should call a perfectly dark one, lying outside the violet, where the incident waves have a length of less than $\frac{1}{64000}$ inch.

As Lord Salisbury said at Oxford, the function of the ether is to undulate, and, in fact, it transports energy from one place to another by wave-motion. Some of its waves, such as those which proceed from an electric-light dynamo, may be thousands of miles in length, others may be shorter than a millionth of an inch, as is perhaps the case with those associated with Professor Röntgen's X-rays; but all, so far as is known, are of essentially the same character, differing from one another only as the billows of the Atlantic differ from the ripples on the surface of a pond. No matter how the disturbance is first set up, whether by the sun, or by a dynamo, or by a warm flat-iron, in every case the ether conveys nothing at all but the energy of wave-motion, and when the waves, encountering some material obstacle which does not reflect them, become quenched, their energy takes another form, and some kind of work is done, or heat is generated in the obstacle.

The whole, or at least the greater part, of the energy given up by the waves is in most cases transformed into heat, but under special circumstances, as, for instance, when the waves fall upon a green leaf or a living eye, a few of them may perform work of an electrical or chemical nature.

The process of the transmission of energy from one body to another by propagation through an intervening medium has long been spoken of as "radiation," and in recent years the same term has been largely employed to denote the energy itself while in the stage of transmission. "Radiation" in the latter sense —meaning ether wave-energy—includes what is often improperly called light. Light, people say, takes about eight minutes in travelling from the sun to the earth. But while it is on its journey it is not light in the true sense of the word; neither does anything of the nature of light ever start from the sun. Light has no more existence in nature outside a living body than the flavour of onions has; both are merely sensations.

If a boy throws a stone which hits you in the face, you feel a pain; but you do not say that it was a pain which left the boy's hand and travelled through space from him to you. The stone, instead of causing pain in a sentient being, might have broken a window, or knocked down an apple. Just so, the same radiation which, when it chances to encounter an eye, produces a certain sensation, will produce a chemical decomposition if it falls upon a cabbage, an electrical effect in a selenium cell, or a heating effect in almost anything. Why, then, should it be specially identified with the sensation?

"Radiation" also includes, and is nearly synonymous with, what is often miscalled radiant heat. After what has been already indicated, I need hardly say that there is no such thing as radiant heat. The truth is that the sun or other hot body generates wave-energy in the ether at the expense of some of its own heat, and any distant substance which absorbs a portion of this energy generally (but not necessarily) acquires an equivalent quantity of heat. The *result* may be exactly the same as if heat left the hot body and travelled across space to the substance; but the *process* is different. It is like sending a sovereign to a friend by a postal order. You part with a sovereign and he receives one, but the piece of paper which goes through the post is not a sovereign. It is strictly correct to say that the sun loses heat by radiation, just as you lose a sovereign by investing it in the purchase of a postal order. But that is not the same thing as saying that the sun radiates heat.

The term "radiation" has the advantage of avoiding any suggestion of the fallacy that there is some essential difference in the nature of the ether-waves which may happen to terminate their respective careers in the production of light or heat or chemical action or something else; but it is, unfortunately, impossible in the present condition of things to use it as freely as one could wish without pedantry, and we must still often speak of light or of heat when radiation would express our meaning with greater accuracy.

Light, then—to use the term unblushingly in its objectionable but well understood sense — has the property of stimulating certain nerves which exist in many living beings, with the result that, in some unknown and probably unknowable manner, a special sensation is called into play—the sensation of luminosity. And in order that the creature may be able not only to perceive light but also to see things, that is, to appreciate the forms of external objects, it is generally provided with an optical apparatus by means of which the incident light is suitably distributed over a large number of independent sensitive elements.

In man and the higher animals the optical apparatus, or eye, consists of a stiff globular shell, having in front an opening provided with a system of lenses, and, at the back of the interior, a delicate perceptive membrane, upon which the transmitted light is received. So much of the light emitted or reflected from an external object as passes through the lenses, is distributed by them in such a manner as to form what is called an "image" upon the membrane, every elementary point of the image receiving the light which issues from a corresponding point of the object, and no other. The contrivance evidently bears a close resemblance to a photographic camera, the sensitive plate or film, upon which the picture is projected, being analogous to the perceptive membrane.

I am not going to attempt a detailed description of the human eye. It will be sufficient to point out briefly some of its principal features as indicated in the annexed diagrammatic section, Fig. 2.

The opening in front of the globe is covered by a slightly protuberant transparent medium C, which is shaped like

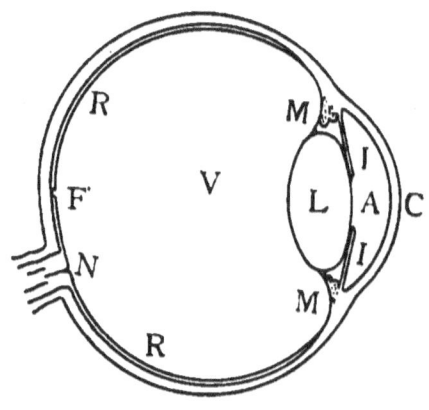

Fig. 2.—*Diagram of the Eye.*

a small watch-glass, and on account of its horn-like structure has been named the *cornea*. The space between the cornea C and the body marked L is filled with a watery liquid A, known as the aqueous humour: this liquid with its curved surfaces constitutes a meniscus lens, convex on the outer side and concave on the inner. Then comes the biconvex *crystalline lens* L, an elastic gelatinous-looking solid, which is easily distorted by pressure. The convexity of this lens can be varied by the action of a surrounding muscle M M, and in this way the focus is adjusted for objects at different distances from the eye. When the muscle is relaxed and the lens in its natural condition, the curvature of its surfaces is such that a sharp image is formed of objects distant about forty feet and upwards. When by an effort of will, the muscle is contracted, the lens becomes more convex, and distinct pictures can thus be focussed of things which are only a few inches away. This process of adjustment by muscular effort is technically known as "accommodation."

The remainder of the globe is filled with the so-called *vitreous body* V, which derives its name from its fancied resemblance to liquid glass: it might perhaps be more properly likened to a thin colourless jelly. The vitreous body plays a part in the refraction of the light.

The perceptive membrane, or *retina* R R, which lines rather more than half the interior of the eye-ball, is an exceedingly complex structure. Though its average thickness is less than $\frac{1}{100}$ inch it is known to consist of nine distinct layers, most of which are marvels of minute intricacy. Of these layers I shall notice only two, the so-called *bacillary layer*, which is in immediate contact with the inner coating of the eye-ball, and the *fibrous layer*, or layer of optic nerve fibres, which is only separated from the vitreous body by a thin protective film.

The bacillary layer (from *bacillum*, a wand) consists of a vast assemblage of little elongated bodies called *rods* and *cones*, which are placed side by side and set perpendicularly to the surfaces of the retina, or in other words, radially to the eye-ball. Let us try to make the arrangement clear by an illustration.

Imagine a small portion of the inner surface of the eye-ball, one-tenth of an inch square, to be magnified 2000 diameters (four million times), and let the enlarged area be represented by the floor of a room 17 feet square. Procure a quantity of cedar pencils, and set them on the floor in an upright position and very close to one another. It will be found that the number of pencils required to fill the space will be about half-a-million. To make the analogy more complete, let some of the pencils be sharpened to a long tapering point at their lower ends, the greater number remaining uncut, just as received from the manufacturers. Neglecting details which are immaterial for our present purpose, we may regard the uncut pencils as representing upon an enormously magnified scale the rods of the retina, and the pointed ones the cones.

The flat upper ends of the pencils may be painted in different uniform colours, and arranged so as to form a large picture in mosaic, and if this is looked at from such a distance that its image on the retina is a tenth of an inch square (which will be the case when the picture is about forty yards away) all possibility of distinguishing the separate elements which compose it will be lost, and the picture will seem to be a perfectly continuous one.

Although the light which enters the eye cannot reach the rods and cones until it has traversed all the other layers of the retina, yet these intervening layers, being transparent, offer little obstruction to its passage, and it can hardly be doubted that the rods and cones are the special organs upon which light exerts its action, the picture focussed upon their ends being in truth an exceedingly fine mosaic.

From every separate element of the

mosaic—from every single rod and cone—there proceeds a slender transparent filament: all these make their way through the intermediate layers of the retina, without, as is believed, any break of functional continuity, and emerge near its internal surface; here they bend over at right angles, and the thousands of filaments form a tangle which lines the inside of the eye like a fine network, and constitute the layer of optic nerve-fibres already referred to.

The filaments, or nerve-fibres, do not however terminate within the eye; they all pass through the hole marked N in the figure, and thence, in the form of a many-stranded cable, constituting the *optic nerve*, they are led to the brain, to which each individual fibre is separately attached. If, therefore, what I have said is true—and, though it has not, I believe, been all rigorously proved, yet the evidence in its support is exceedingly cogent—it follows that every one of the multitude of rods and cones has its own independent line of communication with the brain. The mind, which is mysteriously connected with the brain, is thus afforded the means of localising all the points of luminous excitation relatively to one another, and furnished with data for estimating the form of the object from which the light proceeds.

There are two small regions of the retina which are of special interest. One of them lies just over the opening N where the optic nerve enters. Here it is evident that there can be no rods and cones, their place being wholly occupied by strands of nerve-fibre. Now it is remarkable that this spot is totally insensitive to light.

The other interesting portion is situated opposite the middle of the front opening, and is marked by a small yellow patch, in the centre of which is a depression or pit, which is shown in an exaggerated form at F, and is called the *fovea*. It has been ascertained that the depression is due partly to the absence of the layer of nerve-fibres, which are here bent aside out of their natural course, and partly to a local reduction in the thickness of some of the intermediate retinal layers. This spot, being at the centre of the field of vision, occupies a position of great importance, and the evident purpose of the superficial depression is to allow the light to reach the underlying bacillary layer with as little obstruction as possible. It is noteworthy that the bacillary layer beneath the yellow spot is composed entirely of cones, the rods, which elsewhere are in excess, being altogether wanting.

The only other accessory of the visual apparatus to which I shall refer is the *iris* (I I, Fig. 2), a coloured disk having a central perforation. This can be seen through the cornea and is consequently a very familiar object. The iris serves the same purpose as the stop, or diaphragm, of a photographic lens, its function being to limit and regulate the quantity of light which is admitted into the eye. The size of the central opening, or *pupil*, varies automatically with the intensity of the illumination: in a strong light the opening becomes small; in a feeble light or in darkness it is enlarged.

The pupil also contracts when the eye is focussed upon a near object and dilates when the vision is directed to a distance.

This brief sketch may serve to give some slight idea of the complexity and delicacy of the visual apparatus. Only a few of its more salient features have been touched upon; when our scrutiny is carried into details the complexity becomes bewildering. Even such simple-looking things as the cornea and the vitreous body turn out on close examination to be most elaborately constituted. Much, no doubt, remains to be discovered, and of what has already been investigated much is at present only partially understood.

And yet, though it is true that man is "fearfully and wonderfully made," it is equally true that he is far from perfect; and while there is no structure in the whole human anatomy which exhibits so abundant a profusion of marvels as the eye, there is perhaps none which is marked with imperfections so striking.

Many of its defects are the more striking because they are so obvious, being such as would never be tolerated in optical instruments of human manufacture. In any fairly good camera or telescope or microscope we should expect to find that the lenses were symmetrically figured, free from striæ and properly centred; also that they were achromatic and efficiently corrected for spherical aberration. In the eye not one of these elementary requirements is fulfilled.

The external surface of the lens formed by the aqueous humour and the cornea is not a surface of revolution, such as would be fashioned by a turning lathe or a lens-grinding machine; its curvature is greater in a vertical than in a horizontal direction, and the distinctness of the focussed image is consequently impaired. Again, the crystalline lens is constructed of a number of separate portions which are imperfectly joined together. Striæ occur along the junctions, and the light which traverses them, instead of being uniformly refracted, is scattered irregularly. Moreover the system of lenses is not centred upon a common axis; neither is it achromatic, while the means employed for correcting spherical aberration are inadequate. The purchaser of an optical instrument which turned out to have such faults as these would certainly, as the late Professor Helmholtz remarked, be justified in returning it to the maker and blaming him severely for his carelessness.

I would not, of course, have it believed that scientific men are conceited enough to imagine themselves capable of designing a better eye than is to be found in nature. That would be an absurdity. They are quite ready to admit that there may exist sufficiently good reasons for the undoubted blemishes which have been indicated, as well as for others which will be referred to later. It is indeed well known that the general efficiency of a machine as a whole may often be best secured by the sacrifice of ideal perfection in some of its parts.

With all its anomalies the eye fulfils its proper function very perfectly, and is regarded by those who have studied it most closely with feelings of wonder and humble admiration.*

---

*Possibly the human eye is at present in process of transformation from an inferior type to a different and more perfect one.

CHAPTER II.

COLOUR AND ITS PERCEPTION.

It was explained in the last chapter that we see things through the agency of the light—emitted or reflected—which proceeds from them to the eye, and is suitably distributed over the retina by the action of a system of lenses.

Now the "image" thus formed is not generally perceived as a simple monochromatic one, darker in some parts, lighter in others, like a black and white engraving. It is, in most cases at least, characterised by a variety of colours, the light which comes from different objects, or from different parts of the same object, having the power of exciting different colour sensations. Light which has the property of exciting the sensation of any colour is commonly spoken of as coloured light. The light reflected by a soldier's coat, for example, may be called red light, because when it falls upon the eye it gives rise to a sensation of redness. But it must be understood that this mode of expression is only a convenient abbreviation, for there can, of course, be no objective colour in the light or "radiation" itself.

Wherein, then, does coloured light differ from white? Why do things appear to be variously coloured when illuminated by light which is colourless? And how do coloured lights affect the visual organs so as to evoke appropriate sensations? These are questions—the first two of a physical character, the last partly physiological and partly psychological—which it is now proposed to discuss.

The matter has already been touched upon, though very slightly, in connection with the spectrum. Let us again turn to the spectrum and consider it a little more fully.

It is easily seen that the luminous band contains six principal hues or tones of colour—red, orange, yellow, green, blue, and violet. (See Fig. 1, page 12.) These however merge into one another so gradually that it is impossible to say exactly where any one colour begins and ends. Look, for instance, at the somewhat narrow but very conspicuous stripe of yellow. Towards the right of this stripe the colour gradually becomes greenish-yellow; a little further on it is yellowish-green, and at length, by insensible gradations, a full, pure green is reached.

The six most prominent hues of the spectrum are, in fact, supplemented by an immense multitude of subordinate ones, the total number which the eye can recognise as distinct being not less than a thousand. All the colours that we see in nature, with the exception of the purples (about which I shall say more presently), are here represented, and every single variety of tone in the prismatic scale corresponds with one, and only one, definite wave-length of light.

The source of all these colours is, as we know, a beam of white or colourless light, the constituents of which have been sorted out and arranged so that they fall side by side upon the screen in the order of their several wave-lengths. If, then, these coloured constituents were all mixed together again, it would be reasonable to expect that pure white light would be reproduced.

The experiment has been performed in a great many different ways, several of which were devised by Newton himself, and the result admits of no doubt whatever. The method which I intend to describe is not quite so simple as some others, but it has great advantages in the way of convenient manipulation, and affords the means of demonstrating a number of interesting colour effects in an easily intelligible manner. By the simple operation of moving aside a lens out of the track of the light, we can gather up and thoroughly mix together all the variously coloured rays of the spectrum and cause them to form upon the screen a bright circular patch, which, though due to a mixture of a thousand different hues, is absolutely white. When the lens is replaced, which is done in an instant, the mixture is again analysed into its component parts, and the spectrum reappears.

The arrangement of the apparatus, which is essentially the same as that devised by Captain Abney, and called by him the "colour-patch apparatus," is shown in the annexed diagram (Fig. 3).

The light of an electric lamp A placed inside the lantern is concentrated by the condensing lenses B upon a narrow adjustable slit C. The framework of this slit is attached to one end of a telescope tube, which carries at the other end an achromatic lens D of about 10 inches focus. The rays having been rendered parallel by D are refracted by the prism E; they then pass through a circular opening in the brass plate F to

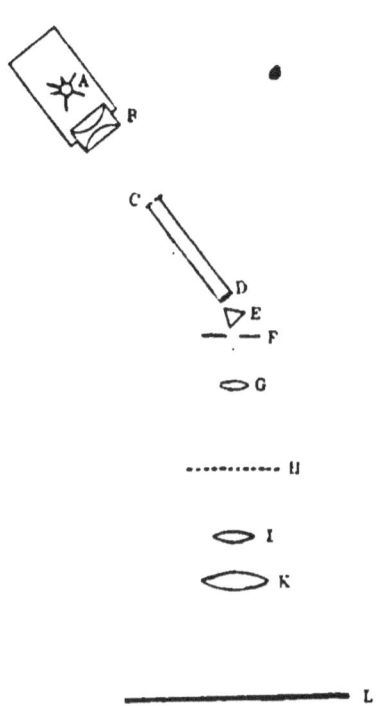

Fig. 3.—*Abney's Colour-patch Apparatus.*

the lens G, the focal length of which is 7 inches, and form a little bright spectrum upon a white card held in a grooved support at H. The card being removed, we place at K a lens having a diameter of $5\frac{1}{2}$ inches and a focal length of 18 inches or more, and adjust it so that a sharply defined image of the hole in the brass plate F is formed upon the distant white screen L. If all the lenses are correctly placed, this image, though formed entirely by the rays which constituted the little spectrum at H, will be perfectly free from colour even around the edge.

If we wish to project upon the screen L an enlarged image of the little spectrum, we have only to use another suitable lens I in conjunction with K: the diameter of that used by myself is $2\frac{3}{4}$ inches, and its focal length $6\frac{1}{2}$ inches. When we have once found by trial the position in which this supplementary lens gives the clearest image* it is easy to arrange a contrivance for removing and replacing it correctly without need of any further adjustment.

This apparatus shows then that ordinary white light may be regarded as a mixture of all the variously coloured lights which occur in the spectrum, the sensation produced when it falls upon the eye being consequently a compound one.

From these and similar experiments the scientific neophyte is not unlikely to draw an erroneous conclusion. White light, he is apt to think, is *always* due to the combined action of rays of every possible wave-length, while coloured light consists of rays of one definite wavelength only. Neither of these inferences

---

* It is sometimes necessary to place the lens I on the other side of K.

would be correct. It is not true that white light necessarily contains rays of all possible wave-lengths : the sensation of whiteness may, as will be shown by and bye, be produced quite as effectively by the combination of only two or three different wave-lengths. Nor is it true that such colours as we see in nature are always due to light of a single wavelength ; light of this kind is indeed rarely met with outside laboratories and lecture rooms. Far more commonly coloured light consists of mixed rays, and like ordinary white light, it may, and generally does, contain all the colours of the spectrum, but in different proportions.

This last assertion is easily proved. By means of a slip of card we may intercept a portion of the little spectrum formed at H (Fig. 3). The dark shadow of the card in the enlarged spectrum on the screen is shown in Fig. 4. It will be noticed that the shadow cuts off a part only of the red, orange, and yellow light, allowing the remainder to pass

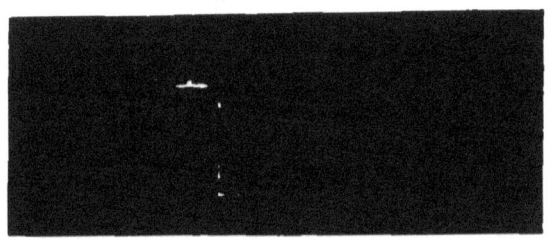

Fig. 4.—*Partially intercepted Spectrum.*

through the projection lenses. There are still rays of every possible wave-length from extreme red to extreme violet, but the proportion of those towards the red end is less than it was before the card was interposed.

If now we remove the lens I (Fig. 3) and so mix the colours of this mutilated
E

spectrum, the bright round patch where the mixed rays fall upon the screen will no longer appear white but greenish-blue. If we transfer the card to the other end of the little spectrum, so as to cause a partial eclipse of the violet, blue, and green rays, the colour of the patch will be changed to orange. If we remove the card altogether, the patch will once more become white.

It follows *a fortiori* that when any portion of the little spectrum is eclipsed totally, instead of only partially, the light from the remainder will appear, when combined, to be coloured. Very beautiful changes of hue are exhibited by the bright patch when a narrow opaque strip, such as the small blade of a pocket knife, is slowly moved along the little spectrum at H, eclipsing different portions of it in succession. The patch first becomes green, then by imperceptible gradations it changes successively to blue, purple, scarlet, orange, yellow, and finally, when the knife has completed its course, all colour disappears and the patch is again white.

We may improve upon this crude experiment, and, after Captain Abney's plan, prepare a number of small cardboard stencils, with openings corresponding to any selected parts of the little spectrum. When a card so prepared is placed at H (Fig. 3) the bright patch upon the screen is formed by the combination of the selected rays, all the others being quenched. We shall find that under these conditions the bright patch is generally, but not always, coloured.

The first diagram in Fig. 5 represents

a blackened card, which allows only the

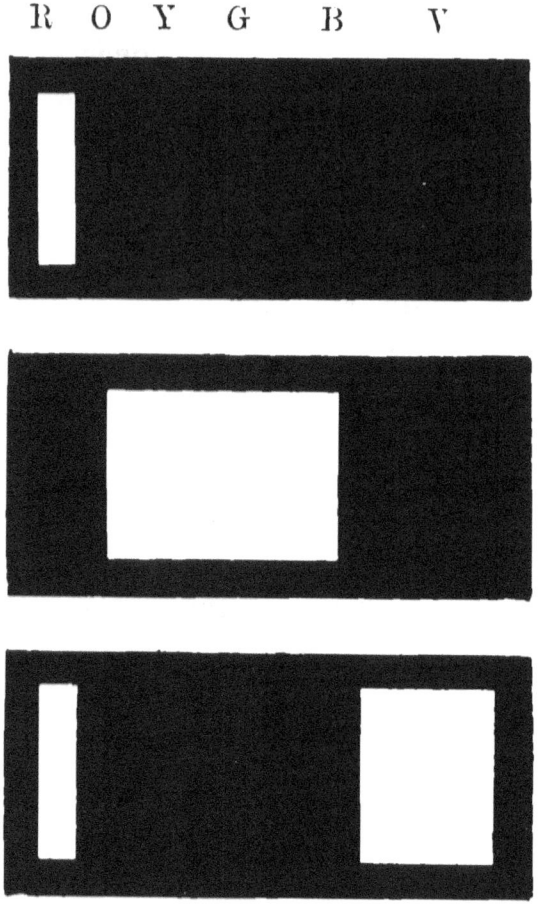

Fig. 5.—*Stencil Cards.*

red and a little of the orange to pass through. When this is inserted in the grooved holder at H, the bright patch immediately turns red. The second diagram shows another, which transmits the middle portion of the spectrum, but blocks the red and the violet at its two ends: with this card the colour of the patch becomes green. The third card has openings for the violet and the red rays: this turns the patch a beautiful purple, a hue which, as already mentioned, is not produced by light of any single wave-length. The purples are mixtures of red and violet or of red and blue.

Now I have in my possession three pieces of glass (or, to be strictly accurate, two pieces of glass and one glass-mounted gelatine film) which, when placed transversely in the beam of light, either at H (Fig. 3) or anywhere else, behave exactly like these three cardboard stencils. The first glass cuts off all the spectrum except the red and part of the orange, just as the first stencil does, though the line of demarcation is not quite so sharp. This is in fact a piece of red glass, or in other words the light that it transmits produces the sensation of red. The second glass, like the second stencil, allows the whole of the spectral rays to pass freely except the red and the violet, which disappear as if they were obstructed by an opaque body. This is a green glass. And the third (which is really a film of gelatine) cuts out the middle of the spectrum but transmits the red and violet ends. The colour of the gelatine is purple.*

---

* It is easy to find specimens of red and green glass suitable for this experiment. The proper kind of purple is not so commonly met with.

The glasses and the gelatine in question act like the cardboard stencils in completely cutting off some of the spectral rays and transmitting others, and they owe their apparent colours to the combined influence which the transmitted rays exert upon the eye. Many other coloured glasses merely weaken some of the rays, without entirely quenching any. A piece of pale yellow glass, for example, when placed in the path of the beam of light from which the spectrum on the screen is formed, simply diminishes the brightness of the blue region and does not wholly quench any of the rays; and again, a common kind of violet-coloured glass enfeebles, but does not quite obliterate, the middle portion of the spectrum.

From such observations as these we infer that the glasses derive their respective colours from the light which falls upon them. The first glass would not appear red if seen in a light which contained no red rays. This is easily proved by an experiment with the colour-patch apparatus. The spectrum being once more combined into a bright white patch (which turns red if the glass is for a moment interposed), let all the red rays and part of the orange be cut off with a suitable stencil. The re-combined light is no longer white but greenish-blue, as is evidenced by the colour of the patch; and nothing that is illuminated by this light can possibly appear red. The piece of red glass, if placed in the beam, will now cast a perfectly black shadow, and a square of bright red paper held in the middle of the patch will look as black as ink. It will be shown later how we may obtain light which, although it appears to the eye to differ in no respect from ordinary white daylight, yet contains no red component, and is consequently as powerless as this greenish-blue light to reveal any red colour in the objects which it illuminates.

If we substitute a stencil which admits only red rays, we shall obtain a beam of light in which no colour but red can be seen. Green and blue glasses when exposed to this light will cast black shadows, while pieces of green and blue paper will become either black or dark grey.

We see then that the colours of transparent objects, like the glasses used in these experiments, are brought out by a process of filtration. Certain of the coloured ingredients of white light are filtered out and quenched inside the glass, and it is to the remaining ingredients which pass through unimpeded that the observed colour is due. The energy of the absorbed rays is not lost of course, for energy, like matter, is indestructible. It is transformed into heat. A coloured glass held in a strong beam of light will in a short time become sensibly warmer than one that is clear and colourless.

In studying colour effects as produced by coloured glasses, we have at the same time been learning how the great majority of natural objects—not only those which are transparent but also those called opaque—become possessed of their colours. For the truth is that few things are perfectly opaque. When white light falls upon a coloured body, it generally penetrates to a small depth below the surface, and in so doing loses by absorption some of its coloured components, just as it does in passing through the pieces of glass. But before it has gone very far—generally much less than a thousandth part of an inch—it has encountered a number of little reflecting surfaces due to optical irregularities, which turn the light back again and compel it to pass a second time through the same thickness of the substance: it thus becomes still more effectively sifted, and on emerging is imbued with a colour due to such of the components as have not been quenched in the course of their double journey through a superficial layer of the substance.

Any coloured rays reflected by an object must necessarily be contained in the light by which the object is seen. The following is a curious experiment illustrating this.

A large bright spectrum is projected upon a screen and in the green or blue portion of it is held a wall poster. The letters and figures upon the paper are seen to stand out boldly as if printed with the blackest ink. But if the poster is moved into the red part of the spectrum, the printing at once disappears as if by magic, and the paper appears perfectly blank. The explanation is that the letters are printed in red ink—they can reflect no light but red. Green or blue light falling upon them is absorbed and quenched, and the letters consequently appear black. On the other hand when the poster is illuminated by the red rays of the spectrum, the letters reflect just as much light as the paper itself, and are therefore indistinguishable from it.

Anything which, when illuminated by a source of white light, reflects all its various components equally and without absorbing a larger proportion of some than of others, appears white or grey. Between white and grey there is no essential difference except in luminosity, or brightness, that is to say, in the quantity of light reflected to the eye, or—to go a step further back—in the amplitude of the ether waves. Under different conditions of illumination any substance which reflects all the rays of the spectrum equally may appear either white or grey, or even black. A snowball can easily be made to look blacker than pitch, and a block of pitch whiter than snow.

It must have struck many of those who have thought about the matter at all as a most remarkable coincidence that sunlight should be white. White light, as we have seen, consists of a mixture of variously-coloured rays in very different and apparently arbitrary proportions, and if these proportions were a little changed the light would no longer be quite colourless. No ordinary artificial light is so exactly white as that of the sun. The light of candles, gas, oil, and electric glow-lamps is yellow; that of the electric arc (when unaffected by atmospheric absorption) is blue, and that of the incandescent gas burner green. It is exceedingly convenient that the light which serves us for the greater part of our waking lives should happen to be just so constituted that it is colourless.

But on a little further reflection it will, I think, appear that this is not the right way to look at the matter. It is precisely because the hue called white is the one which is associated with the light of our sun that we regard whiteness as synonymous with absence of colour. We take sunlight as our standard of neutrality, and anything that reflects it without altering the proportions of its constituents we consider as being colourless.

There can be little doubt that if the sun were purple instead of white, our sentiments as regards these two hues would be interchanged; we should talk quite naturally of " a pure purple, entirely free from any trace of colour," or perhaps describe a lady's costume as being of a "gaudy white."

Even as things are, the standard of neutrality is not quite a hard and fast one. We have a tendency to regard any artificial light which we may happen to be using, as more free from colour than it would turn out to be if compared directly with sunlight. If in the middle of the day we go suddenly into a gas-lit room, we cannot fail to observe how intensely yellow the illumination at first appears; in a few minutes, however, the colour loses its obtrusiveness and we cease to take much notice of it.

The effect may be partly a physiological one, depending upon unequal fatigue of the various perceptive nerves of the retina; but I believe that it is to a large extent due to mental judgment. The standard of whiteness, or colour-zero, can apparently be changed within certain limits in a very short time, and, as we shall see later, this is only one of many instances in which our organs of vision seem to be incapable of recognising a constant standard of reference.

And now let us consider how it comes about that each elementary portion of the retina—at least in its central region—has the power of distinguishing so many hundreds of different hues. It is incredible that every little area of microscopic dimensions should be furnished with such a multitude of independent organs as would be necessary if each of the many colours met with in nature required a separate organ for its perception; and it is not necessary to suppose anything of the kind.

Experiment shows that all the various hues of the spectrum, as well as all (including white) that can be formed from their mixture, may be derived from no more than three distinct colours. There are, in fact, an indefinite number of triads of colours which, in suitable combinations, are capable of producing the sensation of every tone, tint, and shade of colour which the eye of man has ever beheld.

Old-fashioned books, such as an early edition of Ganot's "Physics," tell us that the three "primary" colours are red, yellow, and blue, and that all others are produced by mixtures of these. This was the basis of Sir David Brewster's theory, which attained a very wide popularity, and even at the present time is held as an article of faith among the great majority of intelligent persons who have not paid any special attention to science. But it is not true. A fatal objection to it is the well-ascertained fact that no combination of red, yellow, and blue, or of any two of them, such as blue and yellow, for example, will produce green.

Yet every painter knows that if he mixes blue and yellow pigments together he gets green. That is one of the first things that a child learns when he is allowed to play with a box of water-colours, and no doubt Brewster was misled by the fact.

The truth is, that the colours of all, or almost all, known blue and yellow pigments happen to be composite. An ordinary blue paint reflects not only blue light, but a large quantity of green as well ; while an ordinary yellow paint reflects a large quantity of green light in addition to yellow. When such paints are mixed together, the blue and yellow hues neutralise one another, and only the green, which is common to both, remains.

The spectrum apparatus will make this clearer. Hold a piece of bright blue glass before the slit; the light passing through the glass will be analysed by the prism, and you will see that it really contains almost as much green as blue. If a yellow glass is substituted, not only will yellow light be transmitted, but, as before, a considerable quantity of green. If now both glasses be placed together before the slit, what will happen ? The yellow glass will stop the blue light transmitted by the blue glass, the blue glass will stop the yellow light transmitted by the yellow glass, and only the green light which both glasses have the power of transmitting will pass through unimpeded, forming a band of pure green colour upon the screen.

The combination of simple blue and yellow lights of suitable relative luminosities results in the formation of white or neutral light. If the blue is a little in excess, the combined light will be of a bluish tint; if the yellow is in excess, the combination will have a yellowish tint. It will never contain any trace of green. The combination of simple spectral blue and yellow is easily effected by the colour-patch apparatus, and the result will be found to bear out what has been said.

Since, then, no mixture of red, yellow, and blue, or of any two of them, will produce green, we cannot regard these colours as being, in Brewster's sense of the term, primary ones.

But it is quite possible to find a group of three different hues—and indeed many such groups—which when made to act upon the eye simultaneously and in the right proportions can give rise to the sensation of any colour whatever. Now this experimental fact is obviously suggestive of a possible converse, namely, that almost every colour sensation may in reality be a compound one, the resultant of not more than three simple sensations. Assuming this to be so, it is evident that if each elementary area of the retina were provided with only three suitable colour organs, nothing more would be requisite for the perception of an indefinite number of distinct colours.

Such a hypothesis was first proposed by Thomas Young at the beginning of the present century; but it came before its time and met with no attention until fifty years later, when it was unearthed by the distinguished physicist and physiologist, Helmholtz, who accorded to it his powerful support and modified it in one or two important details.

According to the Young-Helmholtz theory, as it is now called, there are three different kinds of nerve-fibres distributed over the retina. The first, when separately stimulated, produce the sensation of red, the second that of green, and the third that of violet. Light having the same wave-length as the extreme red rays of the spectrum stimulates the red nerve-fibres only; that having the same wave-length as the extreme violet rays stimulates the violet nerve-fibres only. Light of all intermediate wave-lengths, corresponding to the orange, yellow, green, and blue of the spectrum, stimulates all three sets of nerve-fibres at once, but in different

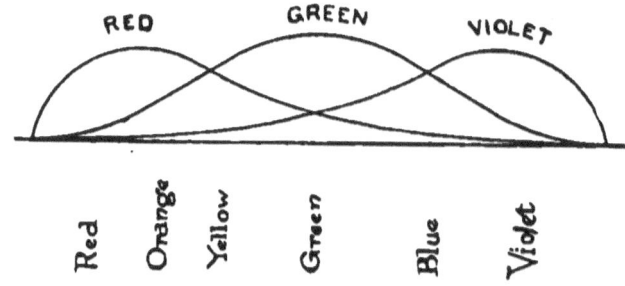

Fig. 6.—*Helmholtz's Curves of Colour Perception.*

degrees. The proportionate stimulation of the red, green, and violet nerves throughout the spectrum is indicated in Fig. 6, which is derived from the rough sketch first given by Helmholtz. The yellow rays of the spectrum, it will be seen, excite the red and green nerves strongly, and the violet feebly; green light excites the green nerves strongly, and the red and violet moderately; while blue light excites the green and violet nerves strongly, and the red feebly.

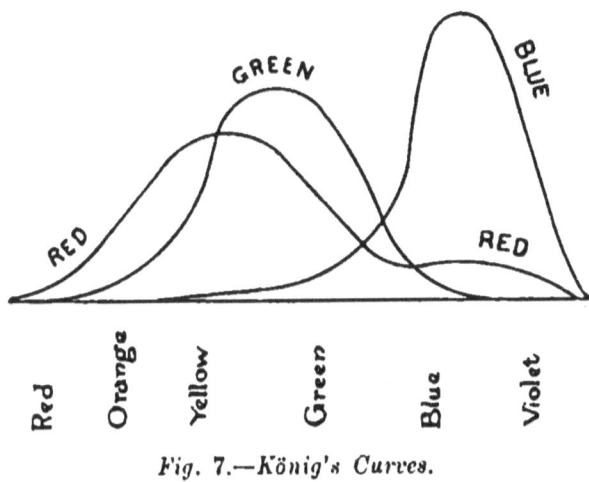

Fig. 7.—*König's Curves.*

Fig. 7 shows another set of curves given more recently by Dr. König as the result of many thousands of experiments made, not only upon persons whose vision was normal, but also upon some who were colour-blind. König found that the equations he obtained were best satisfied by assuming as the normal fundamental sensations a purplish red (not to be found in the spectrum), a green like that of wave-length 5050, and a blue like that of wave-length 4700 approximately, the two latter, however, being purer or more saturated than any actual spectrum colour. But König's curves are not consistent with every class of vision which he examined, and the question as to what are the true fundamental colour-sensations, if such really exist at all, cannot yet be regarded as finally settled.*

---

* Some recent experiments on artificial colour-blindness (Proc. Roy. Soc., Feb., 1898) have led Mr. Burch to the conclusion that there are really *four* fundamental colour-sensations—a red, a green, a blue, and a violet. His results are, however, thought to be capable of a different interpretation.

The Young-Helmholtz theory of colour-vision, whether or not it is destined in the future to be superseded by some other, has at all events proved an invaluable guide in experimental work, and there are very few colour phenomena of which it is not competent to offer a satisfactory explanation. It has at present only one serious rival—the theory of Hering, which, although it seems to be curiously attractive to many physiologists, can hardly be said to present less serious difficulties than that which it seeks to displace. Neither of these competing theories has yet had its fundamental assumptions confirmed by any direct evidence, and the advantage must rest with the one which best accords with the facts of colour vision. In my judgment the older of the two is to be greatly preferred as a useful working hypothesis.

Certain curiosities of vision with which I propose to deal in a future chapter depend upon the properties of what are known as complementary colours. Two colours are said to be complementary to each other when their combination in proper proportions results in the formation of white.

If we produce a compound hue by mixing together the colours of any portion of the spectrum, and a second compound hue by mixing the remainder of the spectrum, it must be evident that these two hues are necessarily complementary, for when they are united they contain together all the elements of the entire spectrum, and therefore appear as white. This may be illustrated with the

COLOUR AND ITS PERCEPTION. 77

aid of the colour-patch apparatus. Place at H (Fig. 3) a cardboard stencil of the form shown in Fig. 8, and focus upon it a little spectrum, the principal hues of which are indicated by the letters R O Y G B V (red, orange, yellow, green, blue,

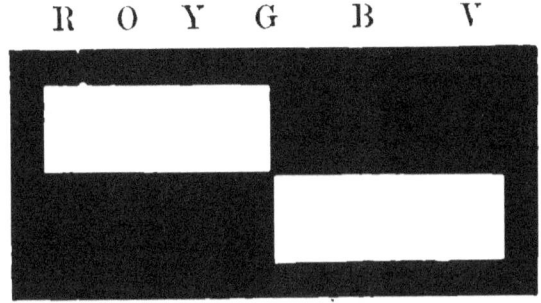

Fig. 8.—Stencil Card for Complementary Colours.

violet). The two oblong apertures in the card should be of exactly the same height, and the card so placed that one aperture may admit rays extending from the red end of the spectrum to about the middle of the green, while the other admits rays from the remainder of the spectrum. If now the lower aperture be covered, only the red, orange, yellow, and part of the green rays will pass through the stencil, and these being combined by the lens K (Fig. 3) will form upon the screen a bright patch, the colour of which will be yellow. If the upper aperture be covered, and the rest of the green, together with the blue and violet rays, allowed to pass through the other, the colour of the patch will become blue; and if both apertures be uncovered at the same time, rays from the whole length of the spectrum will pass through the stencil, and the patch will, of course, turn white. The yellow and the blue which were compounded from the two portions of the spectrum are, therefore, in accordance with the definition, complementary colours.

In a similar manner by dividing the spectrum into any two portions whatever —as, for example, by the complicated stencil shown in Fig. 9—we can obtain an indefinite number of pairs of complementary colours.

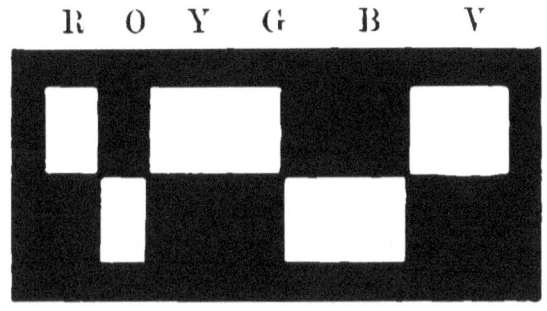

Fig. 9.—*Stencil Card for Complementary Colours.*

But it is by no means indispensable that both or either of a pair of complementary colours should be compound. To prove this, two strips of card with narrow vertical openings A and B are prepared as shown in Fig. 10. The cards are placed one above the other and

80   CURIOSITIES OF LIGHT AND SIGHT.

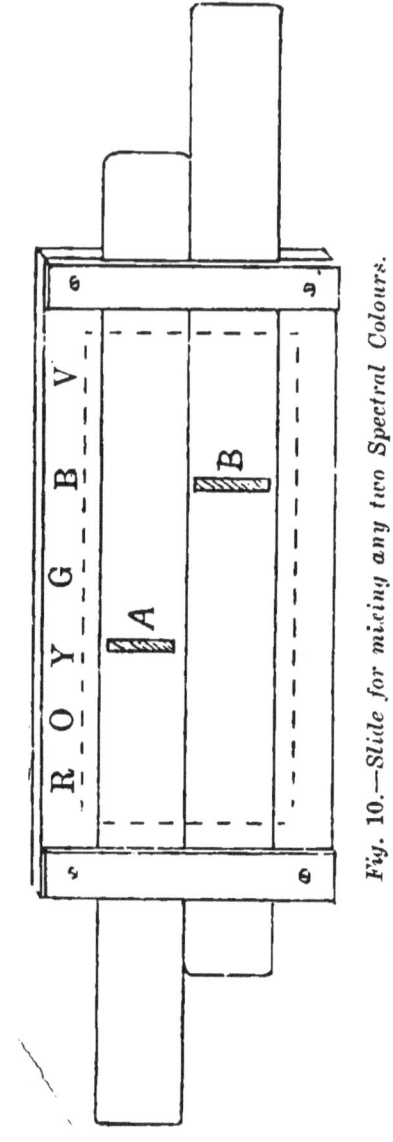

Fig. 10.—Slide for mixing any two Spectral Colours.

can be slipped in a horizontal direction, so that the narrow openings can be brought into any desired part of the spectrum which is indicated in outline by the dotted oblong.

Bring the opening A of the upper card into the yellow of the spectrum and the opening B of the lower card into the blue. The bright patch formed upon the screen will then be illuminated by simple blue and yellow rays; yet it will be white—not green, as it would be if Brewster's theory were correct. If upon the first trial the white should not be absolutely pure, it can easily be made so by partially covering either A or B—the first if the white is yellowish, the second if it is bluish. Simple spectral blue and yellow are therefore no less truly complementary colours than are the compound hues formed when the spectrum is divided into two parts.

It is noticeable, however, that the white light resulting from the combination of blue and yellow, though it cannot be distinguished by the eye from ordinary white light, is yet possessed of very different properties. Most coloured objects when illuminated by it have their hues greatly altered; a piece of ribbon, for example, which in common light is bright red, will appear when held in the blue-yellow light to be of a dark slate colour, almost black.

If the opening A is placed in any part whatever of the spectrum except the green, it will always be possible, by moving B backwards or forwards, to find some other part where the colour is complementary to that at A. To green there is no simple complementary; a purple is required, which is not found in the spectrum, but may be formed by combining small portions of spectral blue and red. For studying mixtures of three simple colours, a third slide may be added to the two shown in Fig. 10.

The following little table gives the principal pairs of complementary colours.

TABLE OF COMPLEMENTARY COLOURS.

Red	Greenish-blue
Orange	Sky-blue
Yellow	Blue
Greenish-yellow ...	Violet
Green	Purple

CHAPTER III.

SOME OPTICAL DEFECTS OF THE EYE.

MORE than one reference has been made to the fact that the sense of sight, even in its best normal condition, is characterised by certain defects and anomalies. Some of these arise directly from causes inherent in the design or structure of the eye itself, and may be broadly classified as physical; others are of psychological origin, and result from the erroneous interpretations placed by the mind upon the phenomena presented to it through the medium of the optic nerve and the brain.

Among the numerous physical defects

of the eye none is more remarkable than the absence of means for properly correcting chromatic aberration. This defect is remarkable because it appears—at least to those who are without actual experience in the manufacture of eyes—to be one which might very easily have been avoided. So far as a mere theorist can judge, an achromatic arrangement of lenses would have been just as simple and just as cheap (if I may use the term) as the arrangement with which we find ourselves provided. It is true that we manage to go through life very well with our uncorrected lenses, and indeed it is hardly possible by ordinary observation to detect any evidence of the imperfection. Yet its existence in a glaring degree is undoubted, and can be readily demonstrated by a great variety of methods. The conclusion is inevitable that with achromatic eyes our vision would be improved, but whether there may not possibly exist reasons why such an improvement could only be achieved at a disproportionately high cost is a question which cannot at present be answered.

Without going into matters which are dealt with in every elementary text book of optics or general physics, it may be desirable to explain shortly what is meant by the terms chromatic aberration, and achromatism.

Let L L, Fig. 11, represent in section a circular convex lens, and P a luminous point, which is most conveniently supposed to be situated on the axis of the lens. Imagine P to be surrounded in the first instance by a glass shade which transmits only monochromatic red light. So much of the light from P as falls upon the lens will be refracted to a point at the conjugate focus F, and after passing this point will diverge again; the refracted light rays will, in fact, form a double cone, of which F is

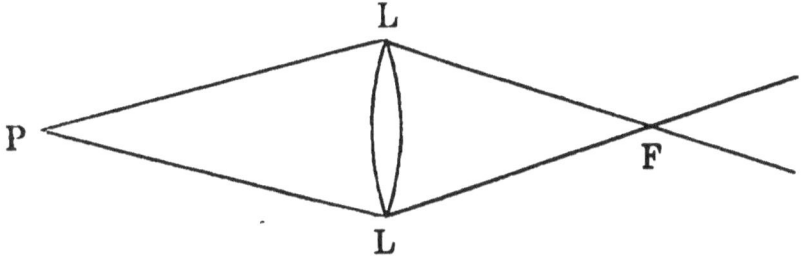

Fig. 11.—*Refraction of monochromatic Light by a lens.*

the apex. If a white screen be held at F, there will be focussed upon it a small clearly-defined image of the luminous point. If, however, the screen be moved nearer to or further from the lens, it will cut the cone of light, and the image will then no longer appear as a point, but as a circular red disk, which will be larger the greater the distance of the screen from F. Such a disk is known as a "diffusion circle."

Suppose now that we substitute for the red glass, surrounding the source of light, a purple one capable of transmitting not only red rays but violet as well. The lens will cause both the red and the violet rays which pass through it to converge; but since the violet rays are more refrangible — more easily refracted or bent aside out of their straight course — than the red, there will now be two double cones, as shown in Fig. 12, where the contours of the red cones are represented by solid lines and those of the violet by dots.

The focus of the red rays will as before be at F, but that of the violet will be nearer to the lens, as at H, and this being so, it is evident that a well defined image of the purple source of light cannot possibly be formed upon a screen placed anywhere behind the lens.

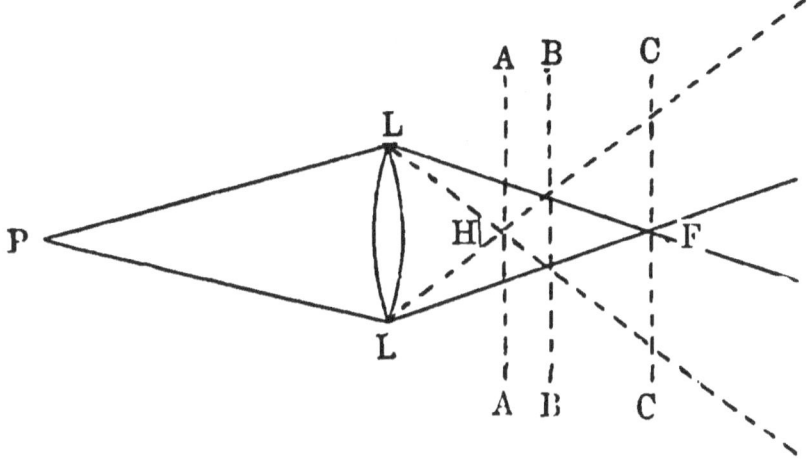

Fig. 12.—*Refraction of dichromatic Light.*

Held in the position indicated by the line C C, where it passes through the focus of the red rays, the screen cuts one of the cones of violet light, and the image at F will appear to be surrounded by a violet halo. Held at A A, the screen evidently receives an image with a red halo round it. Only at B B, in the plane where the surfaces of the red and violet cones cut one another, will it be possible to obtain an image without a coloured border; but here good definition is unattainable, for neither the red nor the violet rays are in focus, and the luminous point is represented by a purple disk or diffusion circle of sensible diameter.

If rays of every possible refrangibility are allowed to fall upon the lens, as is the case when the source of light is not shielded by any coloured glass, there will be formed an indefinite number of pairs of cones, the apices of which will lie along the straight line joining H and F. It is clear that all these cones cannot possibly intersect in a single plane, and consequently no position can be found where the edge of the projected image is perfectly free from colour, though at a certain distance from the lens, where the brightest constituents of the light—namely, the yellow and green—are approximately focussed, the coloured border is least conspicuous, and is of a purple tint, due to the mixture of the red and violet rays.

For these reasons a single glass lens cannot, except with homogeneous light, be made to give a perfectly distinct image of a luminous point, nor of an illuminated object, the surface of which may be regarded as an assemblage of points. Such a lens, therefore, is never employed when good definition is required. The confusion resulting from the unequal refrangibility of the differently coloured rays is said to be due to the chromatic aberration of the lens.

In connection with this matter, the history of physical optics contains an interesting little episode. It occurred to Sir Isaac Newton that although a single lens could never be free from chromatic aberration, yet it might be possible to arrange a so-called achromatic combination of lenses in such a manner as to overcome the defect and bring all the rays issuing from a point, whatever their refrangibility, to one focus. Experiments which he undertook for the purpose of testing the matter led him to form the conclusion that such a result could never be attained, the amount of colour dispersion in all substances being, as he stated, always exactly proportional to that of refraction. For this reason he confidently announced that it was useless to attempt the construction of a really good refracting telescope, and so great was the authority attaching to his name that for many years all efforts in that direction were abandoned.

Nevertheless from time to time certain philosophers ventured to surmise that Newton might perhaps have been mistaken, and the curious thing is that they all based their scepticism upon what they considered the self-evident fact of the achromatism of the eye. The system of lenses in the eye, they argued, being unquestionably achromatic, why should not an equally effective combination be constructed artificially?

At length, more than eighty years after Newton had made and published his fundamental experiments, it occurred to a working optician, John Dollond, that it might be worth while to repeat them, and upon doing so he at once found that Newton was wrong in his facts, the results as recorded by him being in direct opposition to the truth. With proper respect for the memory of a great man it is usual to speak of Newton's observation as a "hasty" one, but if in these days a junior science student were to be guilty of a similar lapse, his conduct would not impossibly be stigmatised as grossly careless.

Having established Newton's error, Dollond found little difficulty in constructing achromatic lenses of very satisfactory quality; telescopes of his manufacture long enjoyed the highest reputation, and the best optical instruments of the present day are the direct offspring of his invention.

Those who entertained the opinion that Newton's conclusion was erroneous were therefore in the right, but it is remarkable that the reason upon which that opinion rested was altogether invalid, for, as I have said, the lenses of the eye are by no means achromatic. Of the many ways in which this can be demonstrated, the following is one of the most impressive.

Let a long and narrow spectrum of the electric light be projected upon a white screen, the prisms and lenses being carefully arranged in such a manner as to ensure that the upper and lower edges of the spectrum are clearly defined and strictly parallel. To an observer standing close to the screen, the spectrum will present the appearance of a bright particoloured rectangle. But viewed from a distance of a few feet the spectrum will not seem to be rectangular, its upper and lower edges no longer appearing to be parallel, but to diverge, fan-like, towards the blue and violet, as shown in Fig. 13. This is because the violet and some of the blue rays proceeding from an object at a little distance cannot by any effort be focussed upon the retina. They are too much refracted, and the mechanism by which the eye is adjusted is incompetent to diminish the convexity of the lenses sufficiently to enable them to project a clear image. Every point is expanded into a luminous circle, which is the larger the more refrangible the rays, and it is the extension of these diffusion circles beyond the proper

SOME OPTICAL DEFECTS OF THE EYE. 97

boundaries of the image that gives the appearance of increased breadth.

It is a simple matter to counteract the effects of undue convexity by means of a concave lens. If a normal-eyed person, to whom the violet end of the

Fig. 13.—Narrow Spectrum as seen from a distance.

spectrum when seen from a distance appears blurred and widened, will look at it through suitable glasses adapted for short sight, he will at once see it clearly defined and of its proper width.

Let a rectangular patch of white light
H

having about the same dimensions as the rectangular spectrum be now thrown upon the screen. The light reflected from the patch will contain, as before, all the various spectral colours, but they will be mixed or superposed, instead of being spread out side by side. The patch will send forth, among others, can yellow and green rays, which the eye easily focus; it will also send out violet rays, which, as we have shown, cannot be focussed by the unassisted eye. Owing to the existence of diffusion circles there must necessarily be formed upon the retina a violet image larger than the approximately superposed images due to rays of brighter colours. Viewed from a distance therefore the white patch might be expected to exhibit a violet border. Yet it may be confidently asserted that the observer will not be conscious of seeing any such border, for though one actually exists, it is possessed of such comparatively feeble luminosity that it is lost in the glare produced by the brighter rays.

It is, however, possible to cut off these brighter rays by interposing between the projection lantern and the screen a combination of glasses which has been found by trial with a spectroscope to transmit only dark blue and violet light. The rectangle will then be of a blue-violet colour, and when looked at closely, will still be quite clear and sharply defined, but viewed from a little distance it will appear blurred and of an exaggerated size.

Another and perhaps even better way of demonstrating this last effect is to enclose the source of light (which should be a powerful one, such as an arc lamp or limelight) inside a box having a ground-glass window in one side. When the window is covered by the coloured glasses its outline cannot be clearly distinguished unless the observer is near, but if he uses suitable concave spectacles, he will be able to see it quite distinctly, even from a considerable distance.

It is well known that ideas of distance are associated with certain colours. A room gives one the impression of being larger when it is papered or painted a blue-violet colour than when its colouring is red. In the former case the walls seem to retire from the spectator, in the latter to approach him. So too a red spot upon a violet ground appears to be distinctly raised above the surface, while a violet spot upon a red ground appears to be depressed. These phenomena are fully explained by the imperfect achromatism of the eye. When we look at a red object, we have to adjust the crystalline lens by means of the ciliary muscle in exactly the same way as when we look at a near object; in both cases it is necessary to increase the convexity of the lens, and so diminish its focal length, in order to obtain a clear image upon the retina. And again, when we wish to see a blue or violet thing distinctly, the ciliary muscle must be relaxed and the convexity of the lens as far as possible diminished, just as if the gaze were directed to the horizon. We are accustomed to estimate the distances of things largely by the muscular effort required to focus their images, and thus it happens that the colour red comes to be associated in our minds with nearness, and violet with remoteness.

These psychological effects are perfectly well marked even with the impure colours met with in ordinary life, but they are naturally more evident when the colours observed are pure, like those of the spectrum.

A beautiful example is that presented by the pair of short bright spectra formed upon the screen when a double slit is used shaped like the letter V. The gorgeously coloured V seems to stand out in strong relief like a pair of inclined boards, the nearer edges being red, the farther ones violet. (See Fig. 14.)

In many other ways, and with little or no apparatus, any one may easily

SOME OPTICAL DEFECTS OF THE EYE. 103

convince himself that the different constituents of white light are not equally refracted by the lenses of the eye. Look, for instance, at the incandescent filament

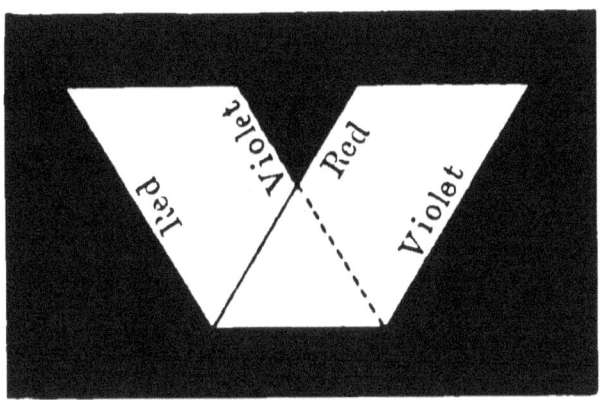

Fig. 14.—*Spectrum formed with V-shaped Slit.*

of an electric lamp through a piece* of common dark blue cobalt glass, which has the property of obstructing the coloured rays corresponding to the middle of the spectrum, while transmitting the red and

---

* Or through several pieces superposed.

the blue. Seen from a distance of only a few inches, the filament appears to be pale blue with a bright red border, the blue rays being perfectly focussed, while the red form diffusion circles. Move some six or eight feet away and look again; the colours will now be reversed, the filament appearing red and the border blue-violet. From a still greater distance—about fifteen or twenty feet—the whole lamp-bulb will seem to be filled with a blue-violet glow, due to large diffusion circles, while the red image of the filament may be even more clearly defined than before. No doubt it is partly owing to the non-achromatism of the eye that distant arc lights always appear to have a yellowish hue, even when the air is quite clear; a considerable proportion of their blue and violet components must necessarily be lost by extensive diffusion.*

Again, look at a sunlit landscape or a printed wall poster through a combination of coloured glasses which will transmit only the violet end of the spectrum. You will find yourself for the time terribly short-sighted, everything appearing blurred and indistinct. But if you resort to the usual corrective for myopia, and put on a pair of concave spectacles, your normal vision will be restored; trees and houses will be seen as clearly as the feebleness of the light transmitted by the coloured glasses will permit, and the letters of the poster will become easily legible.

---

* A violet-coloured haze may sometimes be actually seen around the opal globes of the electric lamps in the streets.

Now, of course, the interposition of coloured glasses does not actually give rise to these blurred images; it merely enables one to detect their existence. Under ordinary conditions they always accompany the clearer images produced by the more luminous rays, and their presence cannot fail to exert a detrimental effect upon the general definition. Such blurs must at least tend to fog the darker portions of the focussed picture, and though we are not distinctly conscious of their existence, it is certain that if they were annulled the acuteness of our vision would be improved.

The diffusion circles produced by the red rays, when the eye is accommodated (as it commonly is) for the yellow and green, are less conspicuous than those due to the most refrangible rays. Yet

I find it impossible to focus a red object, such as the filament of an electric lamp screened by a properly selected deep red glass, when placed at the ordinary distance of distinct vision—some nine or ten inches from the eye — without the aid of a convex lens. In this case one is not too short-sighted but too long-sighted to see the object distinctly; in other words, the lenses of the eye cannot refract the red rays sufficiently to produce well-defined images upon the retina, and the refraction has to be increased by artificial means.

Though, as I have said, it is difficult, or even impossible to detect any trace of a coloured border when looking at a bright object for which the eye is accommodated, it is quite easy to bring such borders into prominence if the object is at a distance a little too great or too small for distinct vision. A very remarkable device for the purpose is one due to von Bezold. This may be illustrated by using a non-achromatic glass lens, such as a common magnifying glass, to project a transparency or lantern-

Fig. 15.—*Bezold's Diagram.*

slide upon which is painted a target-like design, consisting of a series of circular black bands surrounding a circular black spot.* (See Fig. 15.)

Suppose the glass lens to represent the lenses of a gigantic eye (in a definite condition of accommodation) and the screen

---

* A "focus" electric lamp was used in the lantern.

SOME OPTICAL DEFECTS OF THE EYE. 109

the retina. The imaginary eye is looking at the design on the lantern-slide, and when this is at the distance of most distinct vision a fairly well defined image of the target is formed upon the the retinal screen.

Now gradually move the lantern slide towards the lens (or the lens towards the slide), thus bringing it too near for distinct vision. This has the effect of enlarging the diffusion circles formed by the less refrangible rays corresponding to the red end of the spectrum, and at the same time of diminishing those formed by the more refrangible rays corresponding to the violet end. The first result is that the circular dark bands become reddish brown, and the spaces between them bluish. As the distance between the lens and the slide is still further diminished, the tints become more varied and brilliant, until at last there appears a beautiful series of coloured rings around a bright red central spot.

These effects are not produced when the lens employed is an achromatic one; with such a lens the diffusion circles are all enlarged or diminished together, and a to-and-fro movement of the lantern slide (or of the lens) merely affects the definition of the image without causing any perceptible dispersion of colour.

Now it is noteworthy that the chromatic phenomena exhibited with the uncorrected glass lens are quite well shown by the lenses of the eye. It is only necessary to hold the lantern-slide before a bright background and gradually bring it so close to the eye that the design cannot be seen distinctly. The black bands will then appear to turn brown, the white ones blue, and the central spot bright red. The printed diagram (Fig. 15) will itself show the colours if it is held at a distance of four to five inches from one eye in a good light.

One more experiment may be referred to. Look with one eye at a well-lighted page of print, and with a strip of brown paper, held quite near the eye, cover about half the pupil. The black letters will now appear to be bordered with colour—blue towards the apparent edge of the brown paper, orange on the opposite side. If the letters are white on a black ground, as sometimes happens in the case of advertisements, the colours will be interchanged. The cause of the coloured borders will be readily understood from an inspection of the diagram Fig. 12; but it must be remembered that the images on the retina are inverted.

Thus it is proved beyond all question that the lenses of the eye do not form an achromatic combination.

Another peculiarity by which the eye is affected, and which does not occur in optical instruments, is that known as *astigmatism*. The surface of the cornea, which, with the aqueous humour, forms the outer lens, is not often perfectly spherical; generally it is shaped something like the bowl of a spoon, the curvature being greater vertically than horizontally. Rays issuing from a luminous point do not, after refraction by such a lens, cross at a single focus, but

SOME OPTICAL DEFECTS OF THE EYE. 113

along two short straight lines, the one horizontal the other vertical, which are at different distances from the lens; thus a distinct image of a small point cannot anywhere be produced.

A very curious result follows from this

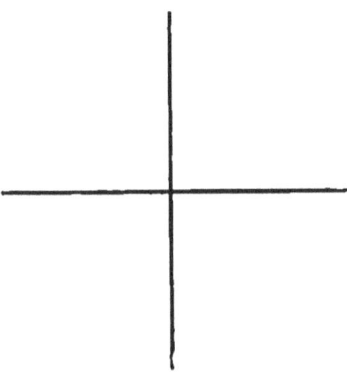

Fig. 16.—*Effect of Astigmatism.*

deformity. If two straight lines are drawn at right angles to each other, as in Fig. 16, it is impossible to see both of them quite clearly at the same time. When the paper is held at a certain short distance from the eye—about eight or nine inches—the horizontal line appears black and well defined, while the other is rather grey and indistinct; at a greater distance the upright line seems to be the blacker. The effect is very well shown by the diagram, Fig. 17. To most persons the lines occupying the middle

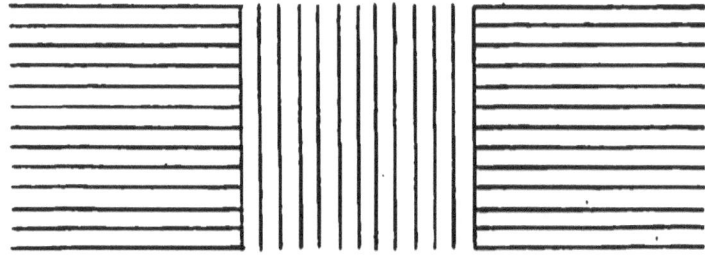

Fig. 17.—*Effect of Astigmatism.*

portion will appear either much blacker or much lighter than those at the two ends, though in fact they are exactly alike. When this form of astigmatism is excessive, it may be corrected by the use of spectacles fitted with cylindrical lenses.

But there is a different kind of astigma-

SOME OPTICAL DEFECTS OF THE EYE. 115

tism—irregular astigmatism it is called—to which every one is more or less a victim, and which cannot be relieved by any artificial appliances. Fortunately it does not often cause much practical inconvenience.

Irregular astigmatism is commonly demonstrated in the following manner. With the point of a fine needle, prick a very small hole in a sheet of tinfoil. Hold up the tinfoil to the light and look at the hole with one eye, the other being closed. Even at the distance of most distinct vision—ten inches or thereabouts,—there will probably be a ragged appearance about the hole, as if it were not perfectly round. But if you bring the tinfoil an inch or two nearer to the eye, the hole will not seem to be even approximately circular; it will assume the form of a little star with five or more distinct rays. The configuration of the star is not generally the same for the right eye as for the left; the rays may differ in number and in relative magnitude, and may be inclined at different angles to the vertical. Fig. 18

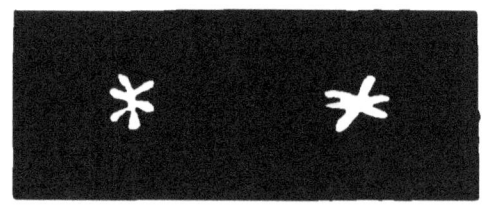

Fig. 18.—*Star-like Images of luminous Point.*

shows the stars as they appear to my two eyes, when the illumination is rather strong.

If several holes are pricked in the tinfoil, each will of course originate a separate star, and all the stars as seen by the same eye will appear to be figured upon the same model, though

SOME OPTICAL DEFECTS OF THE EYE. 117

some may be larger or brighter than others.

There can be no doubt that the stellate form observed in these experiments, as well as that of the stars of heaven themselves (which with perfect vision would be seen simply as luminous points), is a

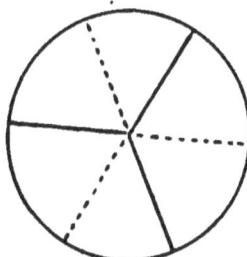

Fig. 19.—*Sutures of crystalline Lens.*

consequence of the singular structure of the crystalline lens of the eye. This does not consist of one uniform homogeneous mass like a glass lens, but of a number of separate portions pieced together radially, as indicated diagrammatically in Fig. 19. In the eye of a newly-born child there are three such portions, and the radial junctions on one side of the lens are not opposite to those on the other, but are intermediate. In the figure the junctions at the front of the lens are represented by continuous lines and those at the back by dots. The number of sutures found in the adult lens is generally greater than six.

But while it is certain that these radial sutures are in some way closely connected with the luminous rays which appear to proceed from a bright point, it must be confessed that no adequate explanation has yet been given of the precise manner in which the phenomenon is brought about. Ophthalmologists seem to have been contented with vague statements about irregular refraction, but what kind of irregularity would sufficiently account for all the facts of observation has never, so far as I know, been exactly determined. The problem can hardly be very difficult of solution, and would, no doubt, readily yield to the joint efforts of a physicist and a physiologist.

The phenomena of irregular astigmatism as exhibited by a normal eye are exceedingly curious, and perhaps I may be allowed to refer briefly to one or two experiments which I have myself made on the subject.*

Light from an enclosed electric lamp of twenty-five candle power was admitted through a circular aperture about $\frac{1}{12}$ inch (2mm.) in diameter perforated in a brass plate; a sheet of ground glass

---

* Proc. Roy. Soc., Jan., 1899.

and another of ruby-red glass were placed behind the aperture. When the little disk of monochromatic light thus formed was looked at through a concave lens of eleven inches focal length from a suitable distance—nearly two feet in

Fig. 20.—*Multiple Images of a luminous Point.*

my own case—it appeared as seven bright round spots upon a less luminous ground. The appearance is represented in a somewhat idealised form in Fig. 20; but the spots were not quite so distinct nor so regularly disposed as there shown, neither was their configuration exactly the same for the right eye as for the left.

On gradually increasing the distance each circumferential spot became at first elongated radially and afterwards split up into two circular ones; at the same time new spots were developed upon the luminous ground, the approximate symmetry of the figure being still retained. Fig. 21 represents a certain stage in this process of expansion. The appearance was happily likened by an observer who repeated the experiment to that of a large unripe blackberry.

As the distance was still further increased, the spots continued to multiply, ultimately becoming very numerous; their arrangement however soon became much less regular, and the definition of most of them less distinct. At about twenty feet there was seen a luminous patch, roughly circular in outline, and covered with irregular speckles; super-

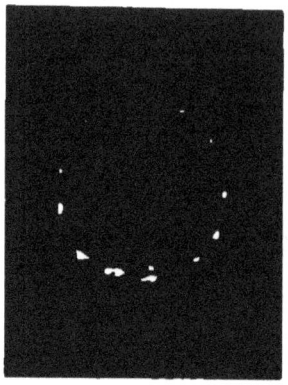

Fig. 21.—*Increased number of Images.*

posed upon this were strings of bright, partially overlapping spots, corresponding apparently to the sutures of the crystalline lens.

When the hole was looked at from a moderate distance through a narrow slit (about ¹⁄₃₀ inch wide) interposed between the eye and the lens, there was seen only a single row of circular spots, which were arranged sinuously, as shown in Fig. 22. A slight movement of the

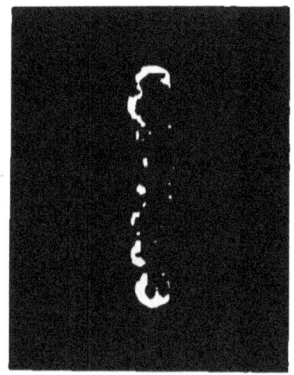

Fig. 22.—*Multiple Images seen through a Slit.*

slit in the direction perpendicular to its length produced a wave-like motion of the circles, suggestive, as pointed out by the excellent observer before referred to, of the wriggling of a caterpillar.

By sufficiently increasing the distance

between the source of light and the eye, as many as twenty-four or twenty-five bright spots might be made to appear in the row, but they could not be counted with any great certainty. At a still longer distance or with a lens of shorter focus (convex or concave) they became less distinct, and finally seemed to be resolved into a multitude of small blurred images—probably several hundreds—which were separated from one another by hazy dark lines.

I thought that the observations might be rendered easier if the source of light had a more distinctive and conspicuous form than that of a simple circle. Some experiments were therefore made with semi-circular and triangular holes, and these were in some respects preferable; but far better results were afterwards

SOME OPTICAL DEFECTS OF THE EYE. 125

obtained by using as a source of light the horse-shoe shaped filament of an electric lamp, screened by a coloured glass. When such a lamp was looked

Fig. 23. — *Images of an electric lamp Filament.*

at through a lens, concave or convex, of about six inches focus, from a distance of a few feet, the roughly oval patch of luminosity formed upon the retina, instead of being a mere ill-defined blur, such as would be produced if the transparent media of the eye were composed of homogeneous substances like glass or water, appeared to be made up of a

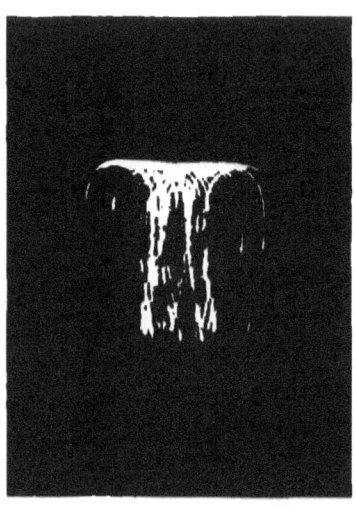

Fig. 24 A.—*Images with horizontal Slit.*

crowd of separate images of the filament, some being brighter than others, as is shown in the diagram Fig. 23.

If a spectroscope slit was interposed between the eye and the lens, and its

SOME OPTICAL DEFECTS OF THE EYE. 127

width suitably adjusted, only a single row of filaments was observed, the appearances with the slit in horizontal, vertical, and intermediate positions being

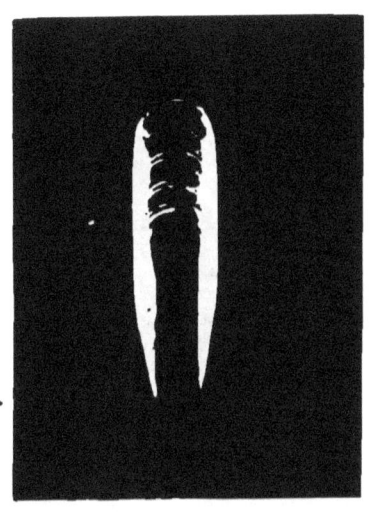

Fig. 24 B.—*Images with vertical Slit.*

as represented in Fig. 24, A, B, C. As before, it was found possible by gradually retiring from the lamp to bring the number of images up to about twenty-five, but attentive examination showed that most of these really consisted of clusters, each composed of perhaps fifteen or twenty confused images of the filament. A stronger

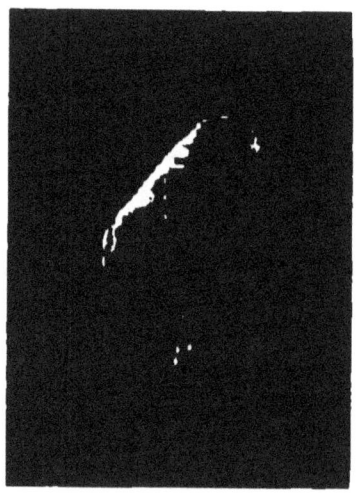

Fig. 24 C.—*Images with oblique Slit.*

lens still further separated the constituents of the clusters, exhibiting a total number of indistinctly seen images which was estimated to amount to nearly five hundred. Assuming the diameter of the pupil of the eye to be one-fifth of an inch, these observations seem to indicate as a cause of the phenomenon some fairly regular anatomical structure, situated in or near the crystalline lens and composed of elements measuring about $\frac{1}{2000}$ inch in length or breadth. Whether the structure which gives rise to these multiple images is to be found in the fibres of the crystalline lens itself, or in the membranes which cover it, is a question upon which I will not venture an opinion.

It is indeed wonderful that an organ affected by peculiarities of which those that have been referred to are merely specimens, should give such well-defined pictures as it does when accommodated for the objects looked at.

CHAPTER IV.

SOME OPTICAL ILLUSIONS.

OPTICAL illusions generally result from the mind's faulty interpretation of phenomena presented to it through the medium of the visual organs. They are of many different kinds, but a large class, which at first sight may seem to have little or nothing in common, arise, I believe, from a single cause, namely, the inability of the mind to form and adhere to a definite scale or standard of measurement.

In specifying quantities and qualities by physical methods, the standards of reference that we employ are invariable.

We may, for example, measure a length by reference to a rule, an interval of time by a clock, a mass or weight by comparison with standardised lumps of metal, and in all such cases—provided that our instruments are good ones and skilfully used—we have every confidence in the constancy and uniformity of our results.

But two lengths, which when tested with the same foot rule are found to be exactly equal, are not necessarily equal in the estimate formed of them by the mind. Look, for instance, at the two lines in Fig. 25. According to the foot rule each of them is just one inch in length, but the mind unhesitatingly pronounces the upright one to be considerably longer than the other; the standard which it applies is not, like a physical one, identical in the two cases. Many other examples might be cited illustrative of the general uncertainty of mental estimates.

The variation of the vague mental standard which we unconsciously employ seems to be governed by a law of

Fig. 25.—Illusion of Length.

very wide if not universal application. Though this law is in itself simple and intelligible enough, it cannot easily be formulated in terms of adequate generality. The best result of my efforts is the following unwieldy statement:—The mental standard which is applied in the estimation of a quality or a condition tends to assimilate itself, as regards the quality or condition in question, to the object or other entity under comparison of which the same (quality or condition) is an attribute.

In plainer but less precise language, there is a disposition to minimise extremes of whatever kind; to underestimate any deviation from a mean or average state of things, and consequently to vary our conception of the mean or standard condition in such a manner that the deviation from it which is presented to our notice in any particular instance may seem to be small rather than large.

Thus, when we look at a thing which impresses us as being long or tall, the mental standard of length is at once increased. It is as if, in making a physical measurement, our foot rule were automatically to add some inches to its length, while still supposed to represent a standard foot: clearly anything measured by means of the augmented rule would seem to contain a fewer number of feet, and, therefore, to be shorter than if the rule had not undergone a change.

It is not an uncommon thing for people visiting Switzerland for the first time to express disappointment at the apparently small height of the mountains. A mountain of 10,000 feet certainly does not seem to be twenty times as lofty as a hill of 500. The fact is that a different scale of measurement is applied in the two cases; though the observer is unaware of it, the mountain is estimated in terms of a larger unit than the hill.

If we mentally compare two adjacent things of unequal length, such as the two straight lines in Fig. 26, there is a tendency to regard the shorter one as longer than it would appear if seen alone, and the longer one as shorter.

Fig. 26.—*Illusion of Length.*

The lower of the two lines in the figure is just twice as long as the other, but it does not look so; each is regarded as differing less than it really does from an imaginary line of intermediate length.

Two divergently oblique lines attached

to the ends of a straight line as at A, Fig. 27, suggest to the mind the idea of lengths greater than that of the straight line itself; the latter, being thought of as comparatively small, is therefore estimated in terms of a smaller unit than would be employed if the attachments

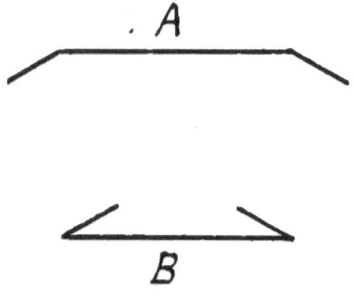

Fig. 27.—*Illusion of Length.*

were absent, and consequently appears longer. If, on the other hand, the attachments are made convergent, as at B, shorter lengths are suggested; the length of the given line is regarded as exceeding an average or mean; the standard applied in estimating it is accordingly increased, and the line is made to seem unduly short. In spite of appearances to the contrary, the two lines A and B are actually of the same length.

By duplicating the attached lines, as shown in Fig. 28, their misleading effect

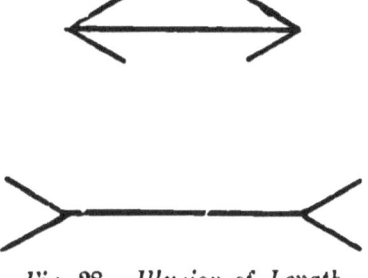

Fig. 28.—*Illusion of Length.*

becomes intensified. Here we have a well-known illusion of which several explanations have been proposed. The fallacy is, I think, sufficiently accounted for by variation of the mental standard, in accordance with the law to which I have called attention.

A number of other paradoxical effects may be referred to the operation of the same law. Fig. 29 shows a curious specimen. At each end of the diagram is a short upright line; exactly in the middle is another; between the middle and the left hand end are inserted several more lines, the space to the right of the middle being left blank.

Fig. 29.—*Illusion of Distance.*

Any one looking casually at the diagram would be inclined to suppose that it was not equally divided by what purports to be the middle line, the left hand portion appearing sensibly longer than the other.

It is not difficult to indicate the

source of the illusion. When we look at the left hand portion we attend to the small subdivisions, and the mental unit becomes correspondingly small; while in the estimation of the portion which is not subdivided a larger unit is applied.

As one more example I may refer to a familiar trap for the unwary. Ask a person to mark upon the wall of a room the height above the floor which he thinks will correspond to that of a gentleman's tall hat. Unless he has been beguiled on a former occasion, he will certainly place the mark several inches too high. Obviously the height of a hat is unconsciously estimated in terms of a smaller standard than that of a room.

The illusion presented by the horizontal and vertical lines in Fig. 25 (p. 132) depends, though a little less directly, upon a similar cause. We habitually apply a larger standard in the estimation of horizontal than of vertical distances, because the horizontal magnitudes to which we are accustomed are upon the whole very much greater than the vertical ones. The heights of houses, towers, spires, trees, or even mountains are insignificant in comparison with the horizontal extension of the earth's surface, and of many things upon it, to which our notice is constantly directed. For this reason, we have come to associate horizontality with greater extension and verticality with less, and, in conformity with our law, a given distance appears longer when reckoned vertically than when reckoned horizontally. Hence the illusion in Fig. 25.

But it is not only in regard to lengths and distances that the law in question holds good; in most, if not all cases in which a psycho-optical estimate is possible, the mental standard is unstable and tends to assimilate itself, as regards the quality or condition to be estimated, to the entity in which the same is manifested. This is true, for example, in judging of an angle of inclination or slope; of a motion in space; of luminous intensity, or of the purity of a colour.

Every cyclist knows how difficult it is to form a correct judgment of the steepness of a hill by merely looking at it. Not only may a slope seem to be greater or less than it really is, but under certain circumstances a dead level sometimes appears as an upward or downward inclination, while a gentle ascent may even be mistaken for a descent, and *vice versa.*

We usually specify a slope by its inclination to a level plane which is parallel to the plane of the horizon, or at right angles to the direction of gravity. At any given spot the level is, physically considered, definite and unalterable. In forming a mental judgment of an inclination, we employ as our standard of reference an imaginary plane which is intended to be identical with the physical level. But our mental plane is not absolutely stable; when we refer a slope to it, we unconsciously give the mental plane a slight tilt, tending to make it parallel with the slope. Hence the inclination of a simple slope, when

SOME OPTICAL ILLUSIONS. 143

misleading complications are absent, is always underestimated.

This may be illustrated by the diagram Fig. 30. If A B represents a truly horizontal line, the slope of the oblique

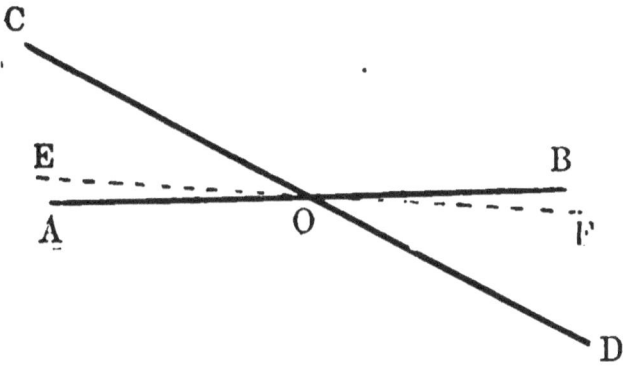

Fig. 30.—*Illusion of Inclination.*

line C D is correctly specified by the angle C O A. But if we have no instrument at hand to fix the level for us, we shall infallibly imagine it to be in some such position as that indicated (in an exaggerated degree) by the dotted line

E F, while the true level A B will appear to slope oppositely to C D.

This class of illusion is remarkably well demonstrated by Zöllner's lines, Fig. 31; the two thick lines which appear to diverge from left to right, are in truth strictly parallel.

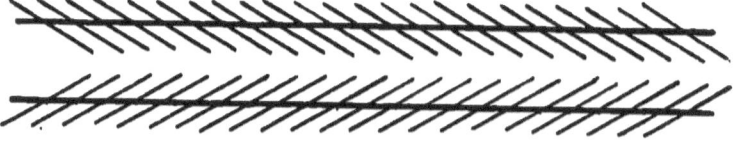

Fig. 31.—*Zöllner's Lines.*

I need not discuss in further detail the various illusions to which a cyclist is subjected when slopes of different inclinations succeed one another: they all follow simply from the same general principle.

A thing is said to be in motion when it is changing its position relatively to the earth, which for all practical purposes may be regarded as motionless. The state, as regards motion, of the earth and anything rigidly attached to it, therefore constitutes the physical zero or standard to which the motion of everything terrestrial is referred. But the corresponding mental standard, especially when it cannot easily be checked by comparison with some stationary object, is liable to deviate from the physical one; it tends in fact to move in the same direction as the moving body which is under observation, and the apparent speed of the body is consequently rather less than it should be.

The influence exerted upon the judgment sometimes even persists for an appreciable period after the exciting cause has ceased to be operative, as when the moving body is lost sight of or has suddenly come to rest; in such cases fixed objects, being compared with the delusive mental standard, appear for a few seconds to be moving in the opposite direction.

I have devised a lantern slide (Fig. 32) by the aid of which this phenomenon may be rendered very evident. In a square plate of metal is cut a vertical slot, which is shaded in the figure; behind the plate is an opaque disk, which, by means of suitable mechanism, can be made to rotate about its centre. The disk has a spiral opening cut in it of the same width as the slot, as indicated by the dotted line. The slide is placed in an optical lantern, and the light passing through the aperture formed where the slot is crossed by the spiral opening, produces a small bright patch upon a white screen hung at a suitable distance from the lantern.

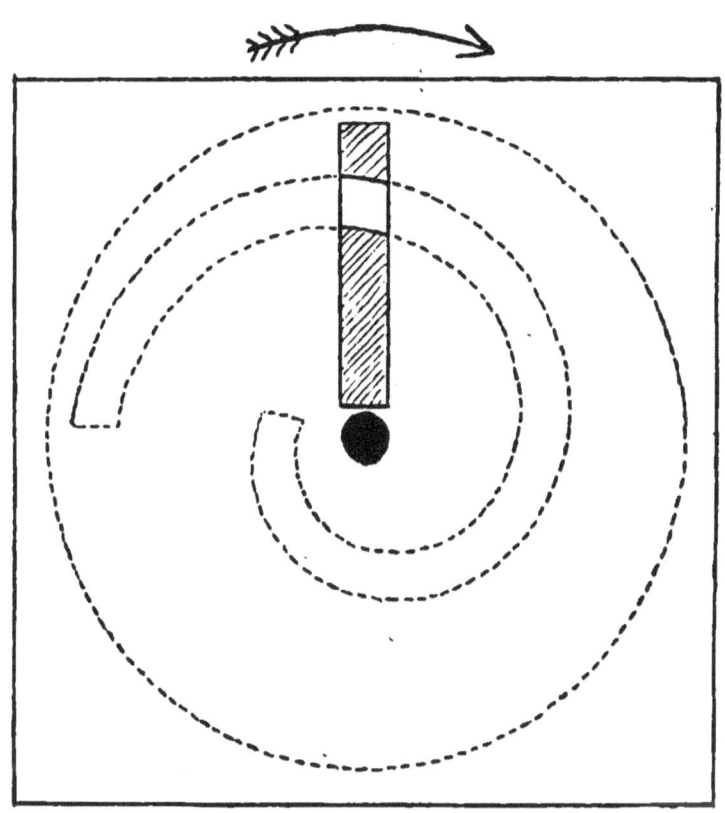

Fig. 32.—*Slide for showing Illusions of Motion.*

When the disk is turned in the direction indicated by the arrow, the bright patch moves upwards and ultimately disappears; but at the moment of its disappearance a fresh patch starts from below, which also moves in the upward direction; thus there is formed upon the screen a continuous succession of ascending bright patches. After these have been observed for about a quarter of a minute, the disk is suddenly stopped, and the persistence of the fallacious mental standard is at once demonstrated. For the bright patch does not appear to be at rest, as it actually is, but to creep steadily downwards, continuing to do so more and more slowly for perhaps as long as ten seconds. The upward motion of the bright patches had led the observer to assume a slower upward motion as the zero, or standard of no motion, and reference of the really stationary patch to this physically false standard induces the illusion that the patch is descending.

This experiment is most successful when the bright patches are projected upon the middle of a large screen. The disk should turn about three times in a second, and the room should be feebly illuminated, but not quite dark.

Fig. 33.—Illusion of Motion.

A very remarkable illusion which no doubt depends upon the same principle as the last, though its form is entirely different, is that to which the diagram Fig. 33 relates. So far as I am aware, it has not before been noticed.

Two intersecting straight lines, the one upright and the other sloping, as shown in the figure, are drawn upon a card. The card is to be held vertically before the eyes at the distance of most distinct vision, and waved up and down through a distance of a few inches. The oblique line will then appear to oscillate transversely, as if it were not rigidly attached to the card.

This is the result of underestimating the speed at which the card is moved. Rather than recognise the true state of things, the mind prefers to accept the suggestion that the upward or downward movement of the point of intersection is in part due to oppositely directed horizontal movements of the lines themselves upon the surface of the card. When the card is descending the vertical line is supposed to slide a little to the right and the oblique line to the left, which would have the effect of lowering their point of intersection independently of the downward movement of the card itself. When the card ascends, these horizontal movements are supposed to be reversed, and the point of intersection consequently raised. The assumption is exactly analogous to that made when an angle of slope is unwittingly minimised.

Another example of the instability of a mental standard occurs in the estimation of luminosity. The luminosity of a bright object, if reckoned in terms of the same unit as that applied in judging of a less bright one, would appear to be greater than it actually does appear, and this quite independently of any effects of fatigue.

152 CURIOSITIES OF LIGHT AND SIGHT.

The fact is well illustrated by a familiar experiment. Fig. 34 is photographed from a transparency made by superposing several different lengths of gelatine film so as to form a series of

Fig. 34.—*Illusion of Luminosity.*

steps. At the right-hand end of the image the light has passed through only one layer of the film; in the next division it has traversed two layers, in the next, three, and in the last, four. The luminosity of each of the four squares into which the oblong is divided is, in a physical sense, quite uniform, but the mental standard of luminosity varies for different parts of the image, increasing or decreasing, as the case may be, not *per saltum*, but smoothly and continuously, with the result that each square looks brighter towards the left than towards the right. The appearance, which is often likened to that presented by a fragment of a fluted column, is equally well shown when the diagram is illuminated instantaneously by an electric spark, and cannot, therefore, be accounted for by retinal fatigue.

If the squares are separated from one another by distinct lines of demarcation, however fine, the standard of luminosity becomes uniform for each square, and the illusion vanishes. This fact sufficiently disposes of the hypothesis which has been advanced to the effect that the phenomenon is due to physiological causes.

I now propose to discuss a curious consequence of the fluctuation of unaided judgment as regards the purity of a colour.

When any colour occupies a predominant place in the field of vision, we are apt to consider it as being less pure, or paler, than we should if it were less conspicuous, our standard of whiteness tending to approximate itself to the colour in question.

For the sake of clearness let us first confine our attention to a definite colour —say red. An absolutely pure red is one that is entirely free from any admixture of white; in proportion as it contains more and more white, the more impure, or in other words, the more pale does it become, until at last all trace of perceptible redness is lost

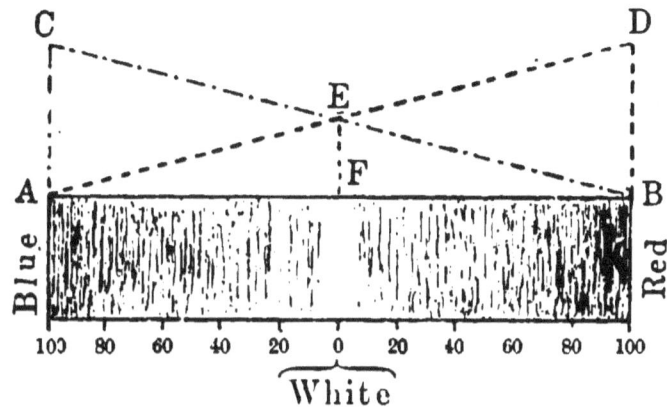

Fig. 35.—*Illusion of Colour.*

and the colour is indistinguishable from white.

A convenient way of picturing the scale of purity is shown in Fig 35. The shaded oblong may be supposed to represent a painted strip of cardboard or paper. At the extreme right hand end the colour is supposed to be absolutely pure red; towards the left the red gradually becomes paler or more dilute, and at the middle of the diagram it has merged into perfect whiteness. The figures 0 to 100 from left to right denote the percentage of free red contained in the mixture at different parts of the scale; the luminosity is supposed to be uniform throughout.

Now the white light with which the red is diluted may be regarded as consisting of two parts, one of which is of exactly the same hue as the pure red itself, and the other an equivalent proportion of the complementary colour, which in the present case will be greenish-blue. The fact therefore really is that, as we pass along the scale from

100 to 0, the *total* quantity of red in the mixture is not reduced to nothing, but only to one half, while at the same time greenish-blue is added in proportions increasing from nought at the extreme right to 50 per cent. of the whole at the middle of the card. The ordinates of the quadrilateral figure E D B F show the proportion of red, and those of the triangle E F B the proportion of greenish-blue, at different parts of the scale.

Regarding the portion of the strip which lies above the point marked 0, as representing the zero of colour—that is, whiteness or greyness, which is essentially the same as whiteness—let us continue the diagram in the negative direction, gradually reducing the quantity of red until it falls from 50 per cent. of the whole at F to nothing at A, and at the same time increasing that of the greenish-blue from 50 per cent. at F to 100 per cent. at A. The resultant hue in the portion of the card between F and A will be greenish-blue, which begins to be perceptible as a very pale tint just to the left of F, and increases in purity as A is approached, at which point the colour will be entirely free from any admixture with white.

We have in the scale thus presented to our imagination a pair of colours, each occupying one-half of the scale, and gradually diminishing in purity towards the middle line; here only, just at the stage where one colour merges into the other, is there no colour at all, and this region represents the fixed physical zero or standard from which is reckoned the purity of a colour corresponding to any other portion of the scale. The completed scale, it will be observed, though originally intended only for the case of red, turns out to be equally serviceable for greenish-blue: if we consider greenish-blue as positive, then the red, being on the other side of zero, must be regarded as negative. Any other possible pairs of complementary colours may be similarly treated.

This device enables us at once to understand the consequence of mentally displacing the zero, while physically the scale remains unchanged. When red is the prevailing colour in the field of vision, we are inclined to consider it unduly pale; in other words we imagine it to be nearer the zero of the scale than is actually the case, and so are led to shift our standard of whiteness from the middle slightly towards the red end of the scale. The new position assigned to white, being a little to the right of the point marked 0 in Fig. 35, is one where, under customary circumstances, the colour would be called pale red. At the same time, an object which is normally white, and is exactly matched at the middle of the scale, would be a little to the left of the imaginary zero, and would consequently appear to be of a greenish-blue tint.

This apparent transformation of white or grey into a decided colour is most striking when the inducing colour is considerably diluted with white or is of feeble luminosity. A small fragment of neutral grey paper, placed upon a much larger piece of a bright red hue, generally appears at the first glance* to be greenish-blue, but if the light is at all strong, only slightly so. If, however, a sheet of white tissue paper is laid over the whole, the greenish-blue tint immediately becomes startlingly distinct, and may even appear more decided than the red itself as seen through the tissue. The same piece of grey paper, when placed upon a green ground, appears rose-coloured, and upon a blue ground, yellow, the effect being always greatly increased by the diluent action of superposed tissue paper.

There seem to be several reasons, partly physical and partly psychological,

---

* After a few seconds' observation the greenish-blue colour often becomes much more intense, but this is an effect of fatigue, with which we are not at present concerned.

why these contrast colours, as they are called, are more pronounced when the colour that calls them into existence either has a somewhat pale tint or is feebly illuminated. Probably the most important is of a purely physical character. The refracting media of the eye are much less perfectly transparent than a good glass lens is; they are sensibly turbid or opalescent, and in consequence of this defect some of the light which falls upon them is irregularly scattered over the retina. If we look at a bright red object with a small white patch upon it, the image of the patch as formed upon the retina is not, physically speaking, perfectly white, but slightly coloured by diffused red light; owing however to the psychological influence to which our attention has been directed, the faint red coloration is not consciously perceived; the same mental displacement of the zero which, when the exciting colour was feeble, led us to regard white (or grey) as bluish-green, now causes what is actually pale red to appear white.

There is no need whatever to assume that the contrast colours with which we have been dealing are of physiological origin and due to an inductive action excited in portions of the retina adjacent to those upon which coloured light falls. On the contrary, it would be a matter for surprise if the case in question presented an exception to the comprehensive law which governs the fluctuation of the mental judgment.

Of the operation of this law I have quoted several very diverse instances, and the number might easily have been increased. Nor is it only in relation to optical phenomena that the law holds good; in its most general form, supplemented it may be in some instances by obvious corollaries, it is applicable to almost every case in which physical attributes of whatever kind are the subject of unassisted mental judgment.

CHAPTER V.

CURIOSITIES OF VISION.

The function of the eye, regarded as an optical instrument, is limited to the formation of luminous images upon the retina. From a purely physical point of view it is a simple enough piece of apparatus, and, as was forcibly pointed out by Helmholtz, it is subject to a number of defects which can be demonstrated by the simplest tests, and which, if they occurred in a shop-bought instrument, would be considered intolerable.

What takes place in the retina itself under luminous excitation, and how the sensation of sight is produced, are questions which belong to the sciences of physiology and psychology; and in the physiological and psychological departments of the visual machinery we meet with an additional host of objectionable peculiarities from which any humanly-constructed apparatus is by the nature of the case free.

Yet in spite of all these drawbacks our eyes do us excellent service, and provided that they are free from actual malformation and have not suffered from injury or disease, we do not often find fault with them. This, however, is not because they are as good as they might be, but because with incessant practice we have acquired a very high degree of skill in their use. If anything is more remarkable than the ease and certainty with which we have learnt to interpret ocular indications, when they are in some sort of conformity with external objects, it is the pertinacity with which we refuse to be misled when our eyes are doing their best to deceive us. In our earliest years we began to find out that we must not believe all we saw; experience gradually taught us that on certain points and under certain circumstances the indications of our organs of vision were uniformly meaningless or fallacious, and we soon discovered that it would save us trouble and add to the comfort of life if we cultivated a habit of completely ignoring all such visual sensations as were of no practical value. In this most of us have been remarkably successful; so much so, that if, from motives of curiosity, or for the sake of scientific experiment, we wish to direct our attention to the sensations in question, and to see things as they actually appear, we can only do so with the greatest difficulty; sometimes, indeed, not at all, unless with the assistance of some specially contrived artifice.

In the present chapter it is proposed to discuss a few of the less familiar vagaries of the visual organs, and to show how they may be demonstrated. Some of the experiments may, it is to be feared, be found rather difficult; success will depend mainly upon the experimentalist's ability to lay aside habit and prejudice, and give close attention to his visual sensations; but it is hardly to be expected that an unskilled person will at the first attempt observe all the phenomena which will be referred to.

Among the most annoying of the eccentricities which characterise the sense of vision is that known as the persistence of impressions. The sensation of sight which is produced by an illuminated object does not cease at the moment when the exciting cause is removed or changed in position; it continues for a period which is generally said to be about a tenth of a second, but may sometimes be much more or less. It is for this reason that we cannot see the details of anything which is in rapid motion, but only an indistinct blur, resulting from the confusion of successive impressions. If a cardboard disk, which is painted in conspicuous black and white sectors is caused to rotate at a sufficiently high speed, the divisions are completely lost sight of, and the whole surface appears to be of a uniformly grey hue. But if the rapidly rotating disk is illuminated by a properly timed series of electric flashes, it looks as if it were at rest, and in spite of the intermittent nature of the light, the black and white sectors can be seen quite continuously, though as a matter of fact the intervals of darkness are very much longer than those of illumination. Persistent impressions of this kind are often spoken of as positive after-images.

There is a very remarkable phenomenon accompanying the formation of positive after-images, especially those following brief illumination, which seems, until comparatively recent times, to have entirely escaped the notice of the most acute observers. It was first observed accidentally by Professor C. A. Young, when he was experimenting with a large electrical machine which had been newly acquired for his laboratory. He noticed that when a powerful Leyden jar discharge took place in a darkened room, any conspicuous object was seen twice at least, with an interval of a trifle less than a quarter of a second, the first time vividly, the second time faintly. Often it was seen a third time, and sometimes, but only with great difficulty, even a fourth time. He gave to this phenomenon the name of recurrent vision; it may perhaps be more appropriately denominated the Young effect.

By means of the powerful machine presented to the Royal Institution by Mr. Wimshurst, used in conjunction with a battery of Leyden jars, the Young effect has been successfully shown to a large assembly. But it is quite easy to demonstrate it on a small scale with any influence machine which will give a spark about an inch long. One of the terminals of the machine should be connected by a wire with the inner coating of a half-pint Leyden jar, the other with the outer coating, and the discharging balls should be set a quarter of an inch apart. The observer's eyes must be shielded from the direct light of the spark by any convenient screen, such as a large book set on end. The best object for the experiment is a sheet of white paper, placed in an upright position a few inches away from the terminals of the machine and exposed to the full light of the discharge.

The room being darkened, let the machine be worked slowly, while the eyes are turned towards the white paper. This will be seen for a moment when the spark passes, and, after a dark interval of about one-fifth of a second, it will make another brief appearance. After a further short interval of darkness, a second recurrent image will often be seen. It may be remarked that the effect is most striking when the eyes are not directed exactly upon the white paper, but above or on one side of it; the proper distance of the paper from the spark-gap should be found by trial.

Under favourable conditions I have observed as many as six or seven reappearances of an object which was illuminated by a single discharge. These followed one another at the usual rate—

about five in a second — and produced a twinkling or quivering effect, closely resembling that attending a flash of lightning which is not directly seen. There can indeed be little doubt that the proverbial quiver of the lightning-flash is in many cases merely an effect of recurrent vision, though sometimes, of course, as has been shown by photographs, the discharge is really multiple.

Some years ago I called attention to a very different method of exhibiting a recurrent image. The apparatus used for the purpose consists of a vacuum tube mounted in the usual way upon a horizontal axis capable of rotation. When the tube is illuminated by a rapid succession of discharges from an induction coil, and is made to rotate very slowly by clockwork (turning once in every two or three seconds), a very curious phenomenon may be noticed. At a distance of a few degrees behind the tube and separated from it by an interval of perfect darkness, comes a ghost. This ghost is in form an exact reproduction of the tube; it is very clearly defined, and though its apparent luminosity is somewhat feeble, it can in most cases be seen without difficulty. The varied colours of the original are, however, absent, the whole of the phantom tube being of a uniform bluish or violet tint. If the rotation is suddenly stopped the ghost still moves steadily on until it reaches the luminous tube, with which it coalesces and so disappears. (See Fig. 36, where the recurrent image is represented by dotted lines.)

More recently a fresh series of experiments were undertaken in connection with the Young effect and certain allied matters, the results being embodied in a communication to the Royal Society (Proc.

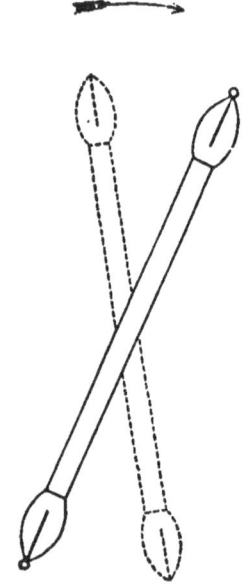

Fig. 36.—*Recurrent Vision demonstrated with a Vacuum Tube.*

Roy. Soc., 1894, vol. 56, p. 132). Among other things an attempt was made to ascertain how far a recurrent image was affected by the colour of the exciting light. With this object two methods of experimenting were employed. In the first, coloured light was obtained by passing white light through coloured glasses; in the second and more perfect series of experiments, the pure coloured light of the spectrum was used. Among other results it was found that, *cæteris paribus,* the recurrent image was much stronger with green light than with any other, and that when the excitation was produced by pure red light, however intense, there was no recurrent image at all.

For a repetition of my first experiment a mechanical lantern slide is required containing a metal disk about three inches in diameter which can be caused to rotate slowly and steadily about its centre. Near the edge of the disk is a small circular aperture. The slide is placed in a limelight lantern, and a bright image of the hole is focussed upon a distant screen, all other light being carefully shut off. When the disk is

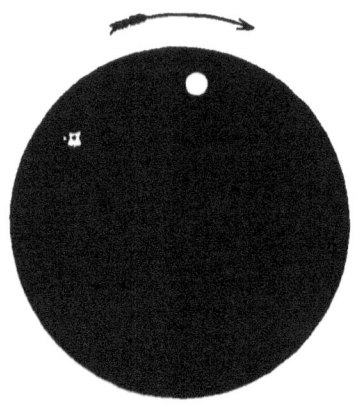

Fig. 37.—*Recurrent Vision with Rotating Disk.*

turned slowly, the spot of light upon the screen goes round and round, and it is generally possible to see at once that the bright primary spot appears to be followed at a short distance by a much feebler spot of a violet colour, which is the recurrent image of the first. (See Fig. 37.) It is essential to keep the direction of the eyes perfectly steady, which is not a very easy thing to do without practice.

If a green glass is placed before the lens, the ghost will be at its best, and should be seen quite clearly and easily, provided that no attempt is made to follow it with the eyes. With an orange glass the ghost becomes less distinctly visible, and its colour generally appears to be greenish-blue, instead of violet as before. When a red glass is substituted, the ghost completely disappears. If the speed of rotation is sufficiently high, the red spot is considerably elongated during its revolution, and its colour ceases to be uniform, the tail assuming a light bluish-pink tint. But however great the speed, no complete separation of the spot into red and pink portions can be effected, and no recurrent image is ever found.

The spectrum method of observation can only be carried out on a small scale, and is not suited for exhibition to an audience. It, however, affords the best means of ascertaining how far the apparent colour of the recurrent image depends upon that of the primary, a matter of some theoretical interest.

The arrangement adopted is shown in the annexed diagram (Fig. 38). L is a lantern containing an oxyhydrogen light or an electric arc lamp, S is an adjustable slit, M a projection lens, P a bisulphide of carbon prism, D a metal plate in the middle of which is a circular aperture 2 millimetres ($\frac{1}{12}$ inch) in diameter. A bright spectrum, 6 or 7 centimetres in length (about 3 inches), is projected upon

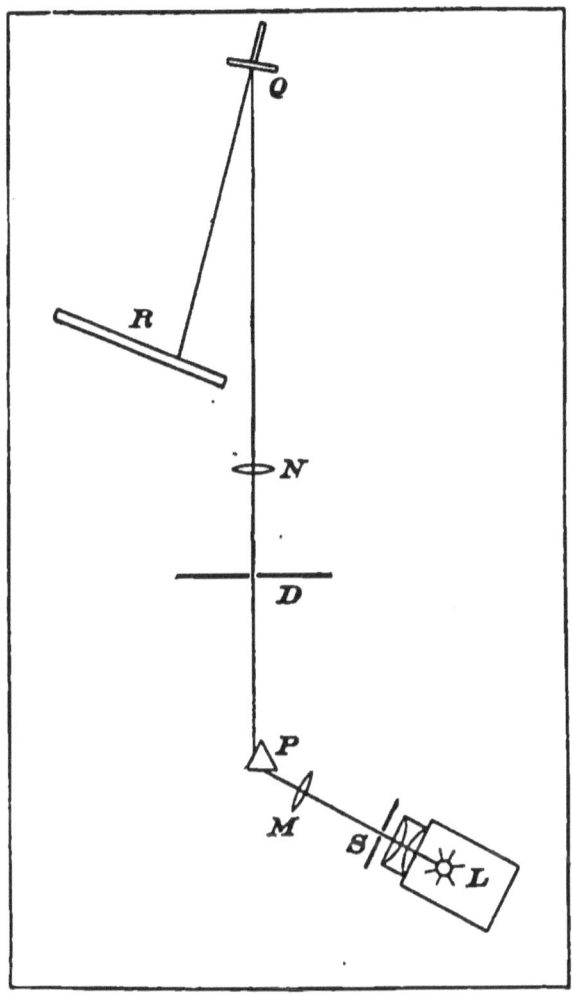

Fig. 38.—*Recurrent Vision with Spectrum.*

this metal plate, and a small selected

portion of it passes through the round hole; thence the coloured light goes through the lens N to the little mirror Q, which reflects it upon the white screen R. By properly adjusting the position of the lens N a sharp monochromatic image of the round hole in the plate D is focussed upon the screen R. To the back of the mirror Q is attached a horizontal arm which is not quite perpendicular to the mirror, its inclination being capable of adjustment. The arm is turned slowly by clock-work, thus causing the coloured spot on the screen to revolve in a circular orbit about 30 centimetres (1 foot) in diameter, its recurrent image following at a short distance behind it. When the mirror turns once in $1\frac{1}{2}$ seconds, this image appears about 50° behind the coloured spot, the corresponding time-interval being about one-fifth of a second.

Using this apparatus, it was found that white light was followed by a violet recurrent image; after blue and green, when the image was brightest, its colour was also violet; after yellow and orange it appeared blue or greenish blue. On the other hand, when a complete spectrum was caused to revolve upon the screen, the whole of its recurrent image from end to end appeared violet; there was no suspicion of blue or greenish-blue at the less refrangible end. For this and other reasons given in the paper it was concluded that the true colour was in all cases really violet, the blue and greenish-blue apparently seen in conjunction with the much brighter yellow and orange of the primary being merely an illusory effect of contrast.

It seems likely, then, that the phenomenon which has been spoken of as recurrent vision, is due principally, if not entirely, to an action of the violet nerve-fibres.

Recurrent vision is, no doubt, generally most conspicuous after a very brief period of retinal illumination, such as was employed in the experiments which we have been discussing; this is evidently due to the fact that the effect is most easily perceived when the sensibility of the retina has not been impaired by fatigue. But by a little effort it may be detected even after very prolonged illumination, and a practised observer can hardly avoid noticing a short flash of bluish light which manifests itself about a quarter of a second after the lights in a room have been suddenly extinguished; the phenomenon forces itself upon my attention almost every night when I turn off the electric lights. It need hardly be pointed out that it represents only a transient phase of the well known positive after-image, and it had even been observed in a vague and uncertain sort of way long before the date of Professor Young's experiment. Helmholtz, for example, mentions the case of a positive after-image which seemed to disappear and then to brighten up again, but he goes on to explain—erroneously, as it turns out—that the seeming disappearance was illusory.

M. Charpentier, of Nancy, whose work in physiological optics is well known, was the first to notice and record a remarkable phenomenon which, in some form or other, must present itself many times daily to every person who is not blind, but which until about seven years ago had been absolutely and universally ignored. The law which is associated with Charpentier's name is this:—When darkness is succeeded by light, the stimulus which the retina at first receives, and which causes the sensation of luminosity, is followed by a brief period of insensibility, resulting in the sensation of momentary darkness. It appears that the dark period begins about one sixtieth of a second after the light has first been admitted to the eye, and lasts for about an equal time. The whole alternation from light to darkness and back again to light is performed so rapidly, that except under certain conditions, which, however, occur frequently enough, it cannot be detected.

The apparatus which Charpentier employed for demonstrating and measuring the duration of this effect is very simple.

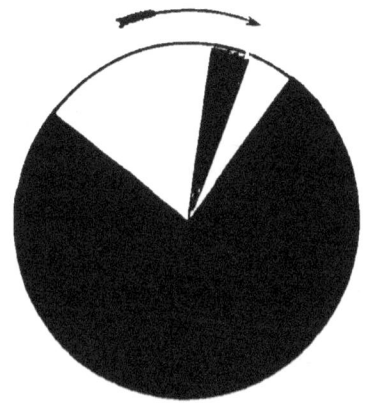

Fig. 39.—*Charpentier's Dark Band.*

It consists of a blackened disk with a white sector, mounted upon an axis. When the disk is illuminated by sunlight and turned rather slowly, the direction of the gaze being fixed upon the centre, there appears upon the white sector, close behind its leading edge, a narrow but quite conspicuous dark band. (See Fig. 39.) The portion of the retina which at any moment is apparently occupied by the dark band, is that upon which the light reflected by the leading edge of the white sector impinged one sixtieth of a second previously.

But no special apparatus is required to show the dark reaction. In Fig. 40 an attempt has been made to illustrate what any one may see if he simply moves his hand between his eyes and the sky or any strongly illuminated white surface. The hand appears to be followed by a dark outline separated from it by a bright interval. The same kind of thing happens, in a more or less marked degree, whenever a dark object moves across a bright background, or a bright object across a dark background.

In order to see the effect distinctly by Charpentier's original method, the illumination must be strong. If, how-

Fig. 40.—*Charpentier's Effect shown with the Hand.*

over, the arrangement is slightly varied, so that transmitted instead of reflected light is made use of, comparatively feeble illumination is sufficient. A very effective way is to turn a small metal disk, having an open sector of about 60°, in front of a sheet of ground or opal glass behind which is a lamp. By an arrangement of this kind upon a larger scale, the effect may easily be rendered visible to an audience. The eyes should not be allowed to follow the disk in its rotation, but should be directed steadily upon the centre.

The acute and educated vision of Charpentier enabled him, even when working with his black and white disk, to detect the existence, under favourable conditions, of a second, and sometimes a third, band of greatly diminished intensity, though he remarks that the observation is a very difficult one. What is probably the same effect can, however, as pointed out in my paper of 1894, be shown quite easily in a different manner.

If a disk with a narrow radial slit, about half a millimetre ($\frac{1}{50}$ inch) wide, is caused to rotate at the rate of about one turn per second in front of a bright background, such as a sheet of ground glass with a lamp behind it, the moving slit assumes the appearance of a fan-shaped luminous patch, the brightness of which diminishes with the distance from the leading edge. And if the eyes are steadily fixed upon the centre of the disk, it will be noticed that this bright image is streaked with a number of dark radial bands, suggestive of the ribs or sticks of a fan. Near the circumference as many as four or five such dark streaks can be distinguished without difficulty; towards the centre they are less conspicuous, owing to the overlapping of the successive images of the slit. The effect is roughly indicated in Fig. 41.

The dark reaction known as the Charpentier effect occurs at the beginning of a period of illumination. There

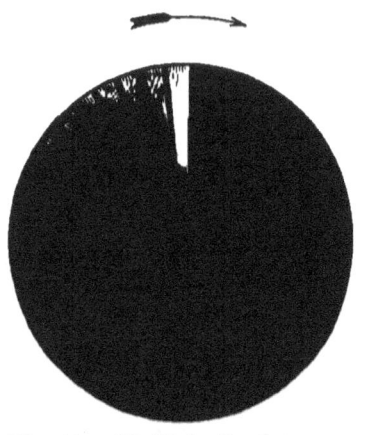

Fig. 41.—*Multiple Dark Bands.*

is also a dark reaction of very short duration at the end of a period of illumination. It should be explained that, owing to what is called the proper light of the retina, ordinary darkness does not appear absolutely black: even in a dark room on a dark night with the eyes carefully covered, there is always some sensation of luminosity which would be sufficient to show up a really black image if one could be produced. Now the darkness which is experienced after the extinction of a light is for a small fraction of a second more intense than common darkness.

The first mention of this dark reaction perhaps occurs in an article contributed to *Nature* in 1885, in which it was stated that when the current was cut off from an illuminated vacuum tube "the luminous image was almost instantly replaced by a corresponding image which seemed to be intensely black upon a less dark background," and which was estimated to last from a-quarter to a-half second. "Abnormal darkness," it was added, "follows as a reaction after luminosity."

In the Royal Society paper before referred to the point is further discussed, and a method is described by which the

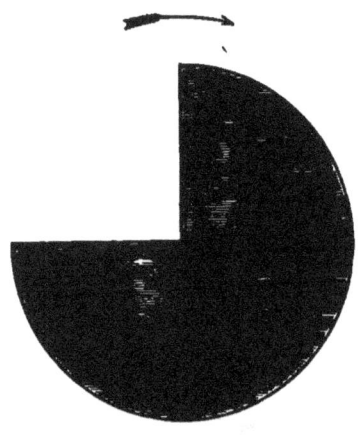

Fig. 42.—*Temporary Insensitiveness of the Eye.*

stage of reaction may be easily exhibited and its duration approximately measured. If a translucent disk, made of stout drawing-paper and having an open sector, is caused to rotate slowly in front of a luminous background, a narrow radial dark band, like a streak of black paint, appears upon the paper very near the edge which follows the open sector. From the space covered by this band when the disk was rotating at a known speed, the duration of the dark reaction was calculated to be about one-fiftieth of a second; my original estimate was therefore an excessive one. The experiment is illustrated in Fig. 42.

One more interesting point should be noticed in the train of visual phenomena which attend a period of illumination. The sensation of luminosity which is excited when light first strikes the eye is for about a sixtieth of a second much more intense than it subsequently becomes. This is shown by the fact, which is obvious enough when once attention has been directed to it, that the bright band, which in the Charpentier disk intervenes between the dark band and the leading edge of the white sector, appears to be much more strongly illuminated than any other portion of the sector.

The complete order of visual phenomena observed when the retina is exposed to the action of light for a limited time may therefore be summed up as follows:—

(1) Immediately upon the impact of the light there is experienced a sensation of luminosity, the intensity of which increases for about one-sixtieth of a second: more rapidly towards the end of that period than at first.

(2) Then ensues a sudden re-action, lasting also for about one-sixtieth of a second, in virtue of which the retina becomes partially insensible to renewed or continued luminous impressions.

These two effects may be repeated in a diminished degree, as often as three or four times.

(3) The stage of fluctuation is succeeded by a sensation of steady luminosity, the intensity of which is, however, considerably below the mean of that experienced during the first one-sixtieth of a second.

(4) After the external light has been shut off, a sensation of diminishing luminosity continues for a short time, and is succeeded by a brief interval of darkness.

(5) Then follows a sudden and clearly-defined sensation of what may be called abnormal darkness—darker than common darkness—which lasts for about one-sixtieth of a second, and is followed by another interval of ordinary darkness.

(6) Finally, in about a fifth of a second after the extinction of the external light, there occurs another transient impression of luminosity, generally violet coloured, after which the uniformity of the darkness remains undisturbed.

Fig. 43, which is copied from my paper, gives a rough diagrammatic representation of the above described chain of sensations. No account is here taken of the comparatively feeble after-images which succeed the recurrent image, and may last for several seconds.

I propose now to say a few words about a curious phenomenon of vision which a short time ago excited considerable interest.

In the year 1895 Mr. C. E. Benham brought out a pretty little toy which he called the Artificial Spectrum Top. It

CURIOSITIES OF VISION. 199

consists of a cardboard disk, one half of

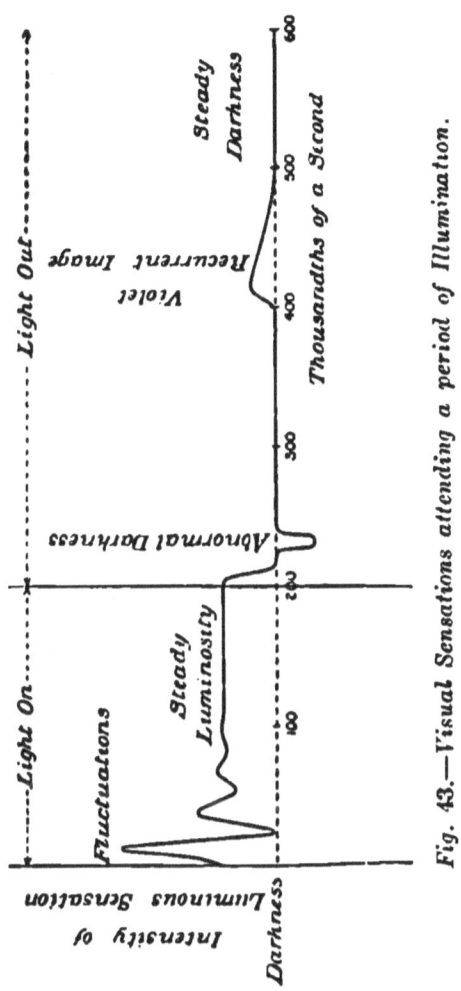

Fig. 43.—Visual Sensations attending a period of Illumination.

which is painted black, while on the

other half are drawn four successive groups of curved black lines at different distances from the centre, as shown

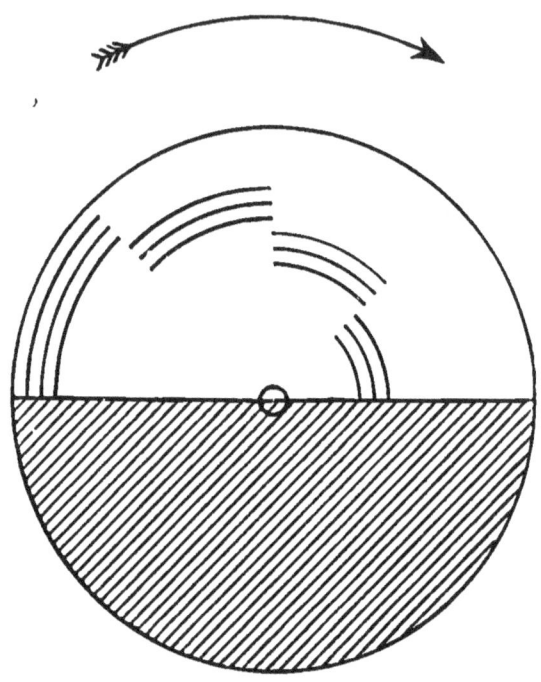

Fig. 44.—Benham's Top.

in Fig. 44. When the disk rotates rather slowly, each group of black lines generally appears to assume a different colour, the nature of which depends upon the speed of the rotation and the intensity and quality of the light. Under the best conditions the inner and outer groups of lines become bright red and dark blue; at the same time the intermediate groups also appear tinted, but the hues which they assume are rather uncertain and difficult to specify. By far the most striking of the colours exhibited by the top is the red, and next to that the blue; this latter is, however, sometimes described as bluish-green.

Some experiments carried out by myself in 1896 (Proc. Roy. Soc., vol. 60, p. 370) seem to indicate pretty clearly the cause of the remarkable bright red colour, and also that of the blue. The more feeble tints of the two intermediate groups of lines perhaps result from similar causes in a modified form, but these have not yet been investigated.

In the red colour we have another striking example of an exceedingly common phenomenon which is habitually disregarded; indeed I can find no record of its ever having been noticed at all. The fact is that whenever a bright image is suddenly formed upon the retina after a period of comparative darkness, this image appears for a short time to be surrounded by a narrow coloured border, the colour, under ordinary conditions of illumination, being red. If the light is very strong, the transient border is greenish-blue, but this colour, as will be explained later, turned out to be merely an after-effect of red. Sometimes, when the object is in motion, both red and blue are seen together.

The observations were first made in the following manner. A blackened zinc plate, in which is a small round hole covered with a piece of thin writing-paper, is fixed over a larger opening in a wooden board; thus we are furnished with a sharply-defined translucent disk, which is surrounded by a perfectly opaque substance. An arrangement is provided for covering the translucent disk with a shutter, which can be opened very rapidly by releasing a strong spring. If this apparatus is held between the eyes and a lamp, and the translucent disk is suddenly disclosed by working the shutter, the disk appears for a short time to be surrounded by a narrow red border. The width of the border is perhaps a millimetre ($\frac{1}{25}$ inch), and the appearance lasts for something like a tenth of a second. Most people are at first quite unable to recognise this effect, the difficulty being, not to see it, but to know that one sees it. Those who have been accustomed to visual observations generally perceive it without any difficulty when they know what to look for, and no doubt it would be very evident to a baby which had not advanced very far in the education of its eyes.

The observation is made rather less difficult by a further device. If the disk is divided into two parts by an opaque strip across the middle, it is clear that each half disk will have its red border, and if the strip is made sufficiently narrow, the red borders along its edges will meet or perhaps overlap, and the whole strip will, for a moment after the shutter is opened, appear red. A disk was thus prepared by gumming across

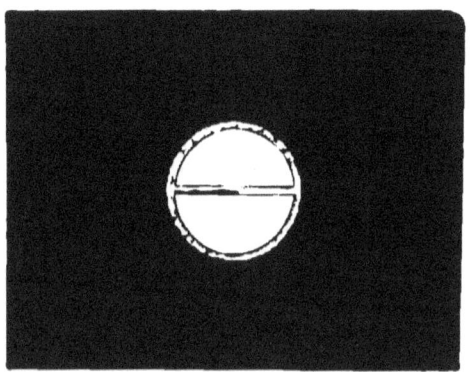

Fig. 45.—Demonstration of Red Borders.

the paper a very narrow strip of tinfoil. The effect produced when such a disk is suddenly exposed is indicated in Fig. 45, the red colour being represented by shading.

A simpler apparatus is, however, quite sufficient for showing the phenomenon,*

---

* See *Nature*, vol. 55, p. 367 (Feb. 18th, 1897).

and with practice one can even acquire the power of seeing it without any artificial aid at all. I have many times noticed flashes of red upon the black letters of a book that I was reading, or upon the edges of the page: bright metallic, or polished objects often show it when they pass across the field of vision in consequence of a movement of the eyes, and it was an accidental observation of this kind which suggested the following easy way of exhibiting the effect experimentally.

An incandescent electric lamp was fixed behind a round hole in a sheet of metal which was attached to a board. The hole was covered with two or three thicknesses of writing paper, making a bright disk of nearly uniform luminosity. When this arrangement was. moved rather quickly either backwards and forwards or round and round in a small circle, the edge of the streak of light thus formed appeared to be bordered with red.

If this experiment is performed with a strong light behind the paper, the streak becomes bordered with greenish-blue instead of red. With an intermediate degree of illumination, both blue and red may be seen together.

Most of the effects that have so far been described were produced by transmitted light, but reflected light will show them equally well. If you place a printed book in front of you near a good lamp and interpose a dark screen before your eyes, then, when the screen is suddenly withdrawn, the printed letters will for a moment appear red, quickly changing to black. Some practice is required before this observation can be made satisfactorily, but by a simple device it is possible to obliterate the image of the letters before the redness has had time to disappear; the colour then becomes quite easily perceptible.

Hold two screens together side by side, a black one and a white one, in such a manner that an open space is left between them. (See Fig. 46.) In the first place let the black screen cover the printing; then quickly move the screens sideways so that the printed letters may be for a moment exposed to view through the gap, stopping the movement as soon as the page is covered by the white screen. During the brief glimpse that will be had of the black letters while the gap is passing over them, they will, if the illumination is suitable, appear to be bright red.

We may go a step further. Cut out a disk of white cardboard, divide it into two equal parts by a straight line

Fig. 46.—*Black and White Screens.*

through the centre, and paint one half black.* At the junction of the black and white portions cut out a gap,

---

* Or, for the best results, use a balanced metal disk covered with black velvet and white paper.

which may conveniently be of the form of a sector of 45°. (See Fig. 47.) Stick a long pin through the centre and hold the arrangement by the pointed end of the pin a few inches above a printed

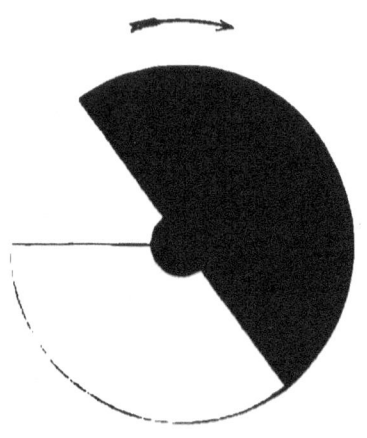

Fig. 47.—*Disk for Red Borders.*

page near a good light. Make the disk spin at the rate of about five or six turns a second by striking the edge with the finger. As before, the letters when seen through the gap will appear red, and persistence will render the repeated impressions almost continuous so long as the rotation is kept up; any one seeing the printing for the first time through the rotating disk would believe that it was done with red ink. Care must be taken that the disk does not cast a shadow upon the page, and that the intensity of the illumination is properly adjusted. I have devised several rather more elaborate contrivances for making the disks rotate at a uniform speed; one of these is shown in Fig. 50.

In none of these experiments does an extended black surface ever appear red, but only black dots or lines. And the lines must not be too thick; if their thickness is much more than a millimetre ($\frac{1}{25}$ inch), the lines, as seen by an observer from the usual distance for reading, do not become red throughout, but only along their edges. The red appearance does not in fact originate in the black lines themselves: these serve merely as a background for showing up the red border which fringes externally the white portions of the paper, and the width of this border does not exceed about one-fifth of a degree. But by employing a sufficiently large disk and selecting designs or letters composed of lines of suitable thickness, the colour effect has been shown to a large audience.

When the disk is turned in the opposite direction, so that the gap is preceded by white and followed by black, the lines of the design appear at first sight to become dark blue instead of red. Attentive observation, however, shows that the apparently blue tint is not formed upon the lines themselves, as the red tint was, but upon the white ground just outside them. This introduces to our notice another border phenomenon, which seems to present itself when a dark patch is suddenly formed on a bright ground, for that is essentially what takes place when the disk is turned the reverse way. I made some attempts to obtain more direct evidence that such a dark patch appeared for a moment to have a blue border, and after some trouble succeeded in doing so.

A circular aperture was cut in a wooden board and covered with white paper; a lamp was placed behind the board, and thus a bright disk was obtained, as in the former experiment. An arrangement was prepared by means of which one half of this bright disk could be suddenly covered by a metal shutter, and it was found that when this was done a narrow blue band appeared on the bright ground just beyond and adjoining the edge of the shutter when it had come to rest. The blue band lasted for about a tenth of a second, and it seemed to disappear by retreating into the black edge of the shutter. The phenomenon is illustrated in Fig. 48, where the shaded band indicates the blue border.

We have then to account, if possible, for the two facts that, in the formation of these transient colour-borders, the red sensation occurs in a portion of the retina which has not itself been exposed to the direct action of light, while the blue occurs in a portion which is steadily illuminated, both colour sensations being referred to localities adjacent to those in which a change of illumination has suddenly taken place. Accepting the Young-Helmholtz theory of colour vision, the effects must, I think, be attributed

Fig. 48.—*Demonstration of Blue Border.*

to a sympathetic affection of the red nerve fibres. When the various nerve fibres occupying a limited portion of the retina are suddenly stimulated by white light (or by any kind of light which contains a red constituent) the immediately surrounding red nerve fibres are for a short period excited sympathetically, while the violet and green fibres are not so excited, or in a much less degree. And again, when light is suddenly cut off from a patch in a bright field, there occurs a sympathetic insensitive reaction in the red fibres just outside the darkened patch, in virtue of which they cease for a moment to respond to the luminous stimulus; the green and violet fibres, by continuing to respond uninterruptedly, give rise to the sensation of a blue border.

It is perhaps desirable to refer briefly to another proposed explanation of the phenomenon, which occurred to myself at an early stage of the investigation, and has since been suggested by many different persons. The explanation in question is of a purely physical character, and depends upon the non-achromatism of the eye.

Without going into details, it will suffice to quote a single experiment

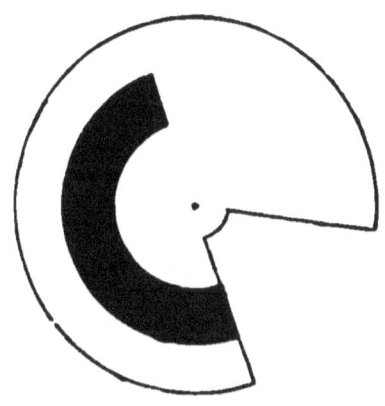

Fig. 49.—*Disk for experiments on the origin of Colour-borders.*

which is of itself fatal to any such theory. Prepare a disk like that shown in Fig. 49, and spin it above a page of printing. The letters beneath the zone which is partly black and partly white will, under the usual conditions, turn red, but those beneath the remainder of the disk will retain their blackness. The demarcation is quite definite, and a single printed word may be made to appear red in the middle and black at its two ends. Now it is, of course, impossible that the lenses of the eye should be perfectly accommodated for the letters which appear black, and at the same time seriously out of focus for the others. This explanation, therefore, simple and obvious as it may seem, is altogether untenable.

Whether or not the hypothesis which I have suggested is correct in all its details, it is, I think, sufficiently obvious that the red and blue colours of Benham's top are due to exactly the same causes as the colours observed in my own experiments, for the essential conditions are the same in both cases.

The last curiosity which I will notice is connected with the fact already mentioned, that when the illumination is strong, the transient border-colours are nearly reversed, greenish-blue appearing in place of red, and brick-red in place of blue.

I was for a long time quite unable to imagine any reasonably probable explanation of this circumstance, but a clue was finally obtained from consideration of the fact that greenish-blue is the complementary colour to red, and in a subsequent memoir (Proc. Roy. Soc., vol. 61, p. 269) some experiments were described which show, as I believe conclusively, that the greenish-blue borders seen in a strong light are simply negative after-images of the usual red one.

These negative after-images are of the familiar kind that are observed after one has gazed for some time at a bright coloured object. If a red "wafer" lying upon a sheet of white or grey paper is looked at steadily for about half a minute, and the gaze is then suddenly transferred to some other part of the paper, a greenish-blue ghost of the wafer will be seen. The portion of the retina upon which the red image at first falls becomes fatigued and partially insensible to red light; it is therefore unable to appreciate the red component of the white light afterwards reflected to it by the paper, and the sensation of the complementary colour consequently predominates; hence the greenish-blue ghost, which is called the negative after-image of the wafer.

The new experiments show that, if a certain condition is fulfilled, the usual prolonged stare becomes unnecessary, a momentary glance sufficing to produce a strong but fugitive after-image. The condition is that the part of the retina where the image is to be formed, shall have been darkened immediately before excitation by the bright object. The retinal nerves, when in darkness, rapidly acquire a state of sensitiveness far exceeding the normal average in the light, but quickly diminishing again under the influence of illumination. This peculiar sensitiveness may, indeed, be both gained and lost in a small fraction of a second, and is therefore very favourable for the rapid generation of negative after-images.

Once more making use of the black

and white screens depicted in Fig. 46, let the black screen first cover the paper upon which the wafer is lying; this will darken a portion of the retina, and render it sensitive. Then let the screens be quickly moved sideways, so that the wafer, after having been seen for a moment through the opening, may be immediately covered by the white screen. A bright but evanescent greenish-blue ghost will succeed the red impression.

But the most curious thing is that if the illumination is strong, and the screens are moved at the proper speed, no trace of red will be seen at all; it will appear exactly as if the actual colour of the wafer seen through the gap were greenish-blue. I am informed that analogous phenomena have been observed in other branches of physiology; a well-defined reaction sometimes occurs when no direct evidence can be detected of the existence of the excitation to which the reaction must be due.

As in the former experiments, the effect may be shown continuously by means of a rotating disk with an open sector. The annexed diagram (Fig. 50) indicates a convenient apparatus for the purpose. The disk is made of thin metal, and properly balanced; the dark portion of the surface is covered with black velvet, and the light portion with unglazed grey or buff paper. It should turn some six or eight times in a second, while its front is well illuminated either by bright diffused daylight or by a powerful lamp. A red card placed behind the rotating disk is made to appear green, a green card pink, and

a blue one yellow, while a black patch

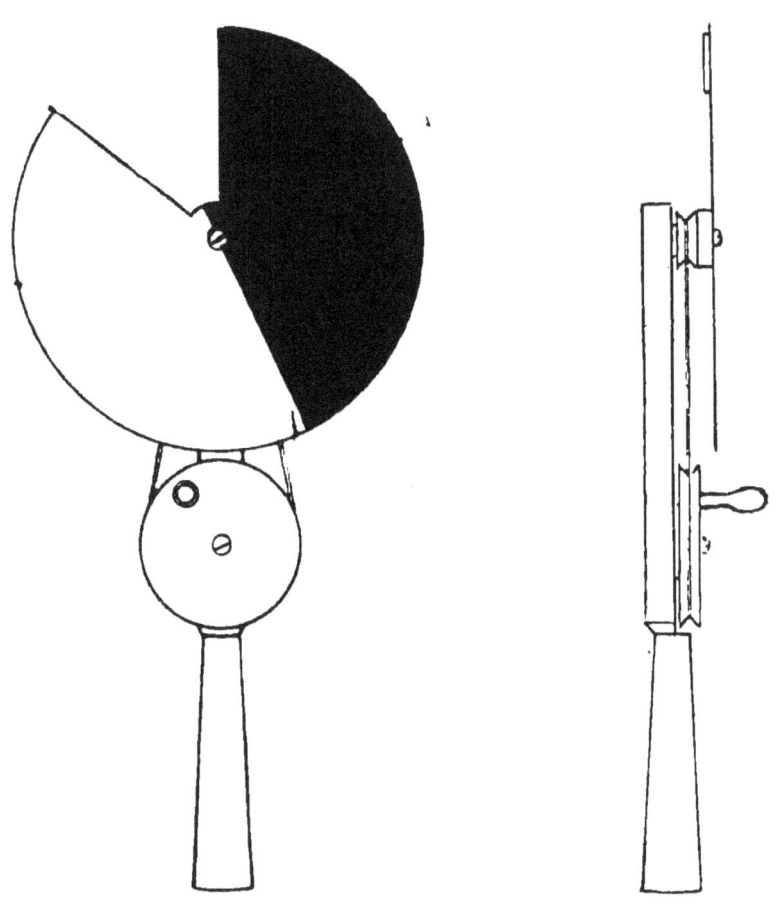

Fig. 50.—*Disk for transforming Colours.*

painted upon a white ground appears

lighter than the ground itself. I have prepared some designs which demonstrate the phenomenon in a very striking manner. One of these is a picture of a lady with indigo-blue hair, an emerald-green face, and a scarlet gown, who is represented as admiring a violet sunflower with purple leaves. Seen through the disk, the lady's tresses appear flaxen, her complexion a delicate pink, and her dress a light peacock-blue; the petals of the sunflower also become yellow, and its foliage green. Other designs show equally remarkable transformations of colour.

I have mentioned only a few among many curious phenomena which have presented themselves in the course of these investigations. It is not improbable that a careful study of the subjective effects produced by intermittent illumination would lead to results tending to clear up several doubtful points in the theory of colour vision.

www.ingramcontent.com/pod-product-compliance
Lightning Source LLC
Chambersburg PA
CBHW021811230426
43669CB00008B/715